T0332417

Cognitive IoT

This book deals with a different research area of cognitive IoT and explains how machine learning algorithms can be applied for cognitive IoT. It deals with applications of cognitive IoT in this pandemic (COVID-19), applications for student performance evaluation, applications for human healthcare for chronic disease prediction, use of wearable sensors and review regarding their energy optimization and how cognitive IoT helps in farming through rainfall prediction and prediction of lake levels.

Features:

- Describes how cognitive IoT is helpful for chronic disease prediction and processing of data gathered from healthcare devices.
- Explains different sensors available for health monitoring.
- Explores application of cognitive IoT in COVID-19 analysis.
- Discusses pertinent and efficient farming applications for sustaining agricultural growth.
- Reviews smart educational aspects such as student response, performance, and behavior and instructor response, performance, and behavior.

This book aims at researchers, professionals and graduate students in Computer Science and Engineering, Computer Applications and Electronics Engineering, and Wireless Communications and Networking.

Intelligent Signal Processing and Data Analysis

In the current era of huge amount of data types and measurement in all sectors and applications, the same requires automated capturing, data analysis and evaluation methods. Consequently, sophisticated intelligent approaches become essential as flexible and powerful tools based on different signal processing algorithms for multiple applications. Intelligent signal processing (ISP) methods are progressively swapping the conventional analog signal processing techniques in several domains, such as speech analysis and processing, biomedical signal analysis, radar and sonar signal processing, and processing, telecommunications, and geophysical signal processing. The main focus of this book series is to find out the new trends and techniques in intelligent signal processing and data analysis leading to scientific breakthroughs in applied applications. Artificial fuzzy logic, deep learning, optimization algorithms, and neural networks are presented for signal processing applications. The main emphasis of this series is to offer both extensiveness of diversity of selected intelligent signal processing and depth in the selected state-of-the-art data analysis techniques for solving real-world problems.

Series Editor:
Nilanjan Dey
Department of Information Technology, Techno India College of Technology, Kolkata, India

Proposals for the series should be sent directly to one of the series editors above, or submitted to:

Chapman & Hall/CRC
Taylor and Francis Group
3 Park Square, Milton Park
Abingdon, OX14 4RN, UK

Cognitive IoT:
Emerging Technology towards Human Wellbeing
J P Patra and Gurudatta Verma

For more information about this series, please visit: https://www.routledge.com/Intelligent-Signal-Processing-and-Data-Analysis/book-series/INSPDA

Cognitive IoT
Emerging Technology towards Human Wellbeing

J P Patra
Gurudatta Verma

CRC Press
Taylor & Francis Group
Boca Raton London New York

CRC Press is an imprint of the
Taylor & Francis Group, an **informa** business

First edition published 2023
by CRC Press
6000 Broken Sound Parkway NW, Suite 300, Boca Raton, FL 33487-2742

and by CRC Press
4 Park Square, Milton Park, Abingdon, Oxon, OX14 4RN

CRC Press is an imprint of Taylor & Francis Group, LLC

© 2023 J P Patra and Gurudatta Verma

ISBN: 9781032315560 (hbk)
ISBN: 9781032315706 (pbk)
ISBN: 9781003310341 (ebk)

DOI: 10.1201/9781003310341

Typeset in Times
by codeMantra

To everyone who made this book possible, I recognize your efforts from the depth of my heart. My parents; my wife Sumitra, my son Yuvraj, colleagues at the Computer Science and Engineering Department, and Institution head – without you people this book wouldn't have been possible. I dedicate this book to all of you.

Dr J. P. Patra

I would like to express our sincere gratitude to everyone who made this book possible. My father R. K. Verma, my mother Dehuti Verma, and especially my wife Khushboo and my daughter Prakriti. I gladly dedicate this publication to you.

Gurudatta Verma

Contents

Preface

This book is intended to present a variety of detailed applications of cognitive IoT and how they might benefit human welfare, as well as how machine learning algorithms can perform cognitive analysis on received data. In this book, we have discussed in detail all the steps of machine learning models such as preprocessing, feature scaling, feature selection, classification, prediction, and validation with their algorithms and applications. Performance indicators such as accuracy, specificity, and sensitivity can be used to validate machine learning models. We've included data preparation, decision tree classifier, KNN classifier, and SVM classifier with MATLAB code in this book so that everyone, from a student to a researcher, may grasp the program and its applications. We just had one goal in mind when we created this book: it should be a boon to students who are working in the field of cognitive IoT or intend to do so in the future.

We primarily constructed this book with the syllabus of many countries in mind so that undergraduate students, postgraduate students, and university research researchers can utilize it for their studies. In Chapter 1, this book mainly focuses on cognitive IoT and its impact on human life. Chapter 2 presents the smart student evaluation mechanism and its implementation using MATLAB. Chapter 3 is focused on chronic disease prediction and heart disease prediction using MATLAB tool. Chapter 4 is about energy-efficient wearables and their applications. Chapter 5 provides brief details about rainfall prediction for effective farming using decision tree classifier, support vector machine, and regression models. Chapter 6 elaborates how cognitive intelligence is applied for the prediction of lake level, as it helps in water conservation and drought-like situation; again, this is very important for human wellbeing, and unsupervised machine learning models are applied for the prediction of lake level.

Acknowledgement

The completion of this book would not have been possible without the participation and assistance of a large number of persons, many of whose names are not listed. Their contributions are heartily welcomed and appreciated. However, we would like to express our deep appreciation and indebtedness, particularly to the following:

Dr Nilanjan Dey, Associate Professor, JIS University, Kolkata; Shri Nishant Tripathi, Chairman (BG), SSIPMT, Raipur; Dr Alok Kumar Jain, Principal, SSIPMT, Raipur; Dr Tirath Prasad Sahu, Assistant Professor, NIT, Raipur; Dr Rekh Ram Janghel, Assistant Professor, NIT, Raipur; Dr Seema Arora, Associate Professor, SSIPMT, Raipur; Dr Ritu Benjamin, Associate Professor, SSIPMT, Raipur; Dr Partha Sarathi Khuntia, Principal, KIST, Bhubaneswar.

We are deeply indebted to our colleagues and friends of the Department of Computer Science and Engineering, Shri Shankaracharya Institute of Professional Management and Technology, Raipur, for their contribution in bringing out this book. With heartfelt thanks, we remember all those people, though not mentioned here, who have played an important role in the success of this book.

Author's Biography

Dr J P Patra is a Professor at Shri Shankaracharya Institute of Professional Management and Technology, Raipur, under Chhattisgarh Swami Vivekanand Technical University, Bhilai, India. He has more than 17 years of experience in research, teaching in the areas of Artificial Intelligence, Analysis and Design of Algorithms, Cryptography, and Network Security. He was acclaimed for being the author of books such as *Analysis and Design of Algorithms* (ISBN-978-93-80674-53-7) and *Performance Improvement of a Dynamic System Using Soft Computing Approaches* (ISBN: 978-3-659-82968-0), and has published more than 51 papers in SCOPUS, Web of Science, and UGC-CARE listed journals. He has published and granted Indian/Australian patents. He has contributed to book chapters, published by Elsevier, Springer, and IGI Global. He is associated with AICTE-IDEA LAB, IIT Bombay, and IIT Kharagpur as a coordinator. He is on the editorial board and reviewer board of four leading international journals. In addition, he is on the Technical Committee Board for several international conferences. He is having a Life Membership of professional bodies such as CSI, ISTE, and QCFI, and he has also served the post of Chairman of the Raipur Chapter for the Computer Society of India, which is India's largest professional body for computer professionals. He has served in various positions in different engineering colleges as Associate Professor and Head. Currently, he is working with SSIPMT, Raipur, as Professor and Head of the Department of Computer Science and Engineering.

Mr Gurudatta Verma is Assistant Professor at Shri Shankaracharya Institute of Professional Management and Technology, Raipur, under Chhattisgarh Swami Vivekanand Technical University, Bhilai, India. He has more than 12 years of experience in research, teaching in the areas of parallel processing and machine learning. He has published more than 15 papers in SCOPUS, Web of Science, and UGC-CARE listed journals. He has published and granted Indian/Australian patents. He has contributed to book chapters published by Elsevier, Springer, and IGI Global.

Cognitive Internet of Things and Its Impact on Human Life

1

1.1 INTRODUCTION TO COGNITIVE INTERNET OF THINGS

The Internet of Things (IoT), first invented by Kevin Ashton as the name of a seminar in 1999 [1], stands for a technological breakthrough getting us in brand-new omnipresent connectivity, information technology, and communication era. Development in the field of IoT is determined by the dynamic technological innovations in various fields, starting from wireless sensors to nanotechnology [2]. In connection with these groundbreaking advancements to grow after ideas to certain products or requests, during the past decade, we have been witnessing efforts from the academic community, service providers, network administrators, etc. all over the world (see, e.g., the most recent detailed surveys in [3]–[5]). Technologically, much of the attention is concentrated on facets, for example, communication, information processing, connection to the Internet, etc., which are extremely important topics. Although we contend that without a comprehensive cognitive ability, IoT is the same as an uncomfortable stegosaurus: all brawn, no brains. To achieve its possibilities and deal with ever-increasing trials, there is a need to take cognitive expertise into account and embolden IoT with high-level intelligence.

Furthermore, the healthcare sector has established itself as one of the key industries with immense demands. In addition to providing sick people with

DOI: 10.1201/9781003310341-1

vital services, this business is likewise producing substantial income for the government as well as the private sector. The smart healthcare industry has also recently seen competition between various healthcare providers in providing mature and older services and devices with high accuracy and reliability and low cost. Therefore, the integration of IoT – cloud – within health care has recently been the focus of much research. Many types of IoT devices designed for health care include smart wearable devices, such as blood pressure devices, portable insulin syringes, stress monitoring devices, weight tracking and standard fitness devices, hearing aids, and electroencephalogram (EEG) and electrocardiogram monitors [6]. Although healthcare data, such as EEG, is naturally complex, we have made many technological advances in the field of big data analytics and cloud computing to manage the complexity of such data and provide the processing power and storage capacity required for that information. However, many IoT-connected devices with sensors and a wide range of multimedia, health care, and communications make it difficult to create a smart healthcare framework that can cater to the needs of all stakeholders in a smart city environment. However, the idea of integrating smart IoT – the cloud – is impossible without intelligence like the human brain. With great detail and its real-time application that comes with this picture, the research community faces several challenges in developing a smart and intelligent IoT-cloud framework, which will be able to make its own decisions. As a result, a computer-assisted computer framework was developed and proposed to convert IoT into IoT (cognitive IoT [CIoT]) brain-enabled brain, which will have a higher level of intelligence [7].

Complex types of concepts and relationships on a scale can be time-consuming and expensive. In addition, many relationships are unknown or obvious in the past, so it is only possible for the machine to automatically analyse big datasets to find patterns. Figure 1.1 shows the application of cognitive IoT in different ways.

Let's have a clear insight into cognitive IoT, using an example. We show the application of cognitive IoT, e.g., our treadmill. Jump on the treadmill. With the smart camera, you can scan and positively identify yourself to us. IoT sensors can measure your body weight, starting by measuring your heart rate. AI pulls medical profile, and the account of the last visit to the cardiologist, etc., to your friends. We are tired. Based on the available data and the output of the pulse, the Artificial Intelligence (AI)-powered brain of a treadmill creates a separate decision as to stop the running belt moves, and you are up to date with your physician and healthcare providers with relevant and sensitive information. While on the treadmill, it could call an ambulance, if needed, to just a detect of a heart attack before it began. Figure 1.2 shows the IoT uses and future uses worldwide.

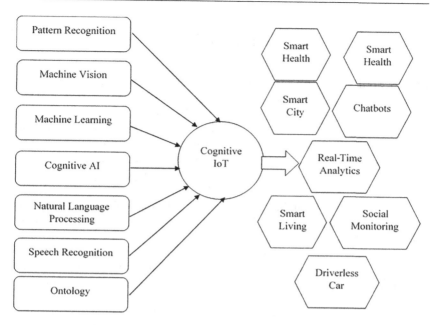

FIGURE 1.1 Cognitive IoT and Functions.

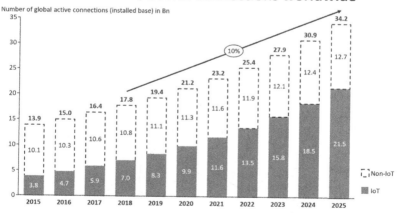

FIGURE 1.2 IoT Worldwide (Source: https://iot-analytics.com/state-of-the-iot-update-q1-q2-2018-number-of-iot-devices-now-7b/).

1.2 IoT

IoT, the word first presented by Kevin Ashton in 1998, is the future of the Internet and pervasive computing. This technical revolution characterizes the future of reachability and connectivity. In IoT, "things" state some object on the face of the globe, whether it is a collaborating device or a non-collaborating dumb object. In the whole world, whatsoever can be part of the Internet. The objects become connecting nodes over the Internet, through data communication channels, mainly from side-to-side Radio Frequency Identification (RFID) tags. IoT includes some Smart Objects (SO) too. Objects those are not only physical units but also digital, which accomplish some jobs for humans and the atmosphere called SO. This is why IoT is hardware as well as a software prototype and also includes interaction and social features on top. Other than portraying the frameworks and things of IoT, various later reviews stressed that most things on the IoT should have the knowledge, in this way are called SO, and are expected fit for being recognized, detecting occasions, communicating with others, and settling on choices independent from anyone else.

Figure 1.3 demonstrates the two separate interaction modes in which smartphones can empower in the IoT. Via direct interaction, the smartphone can question the state of an IoT device in its proximity and then provide a bridge between low-level peer-to-peer protocols, such as Bluetooth or Wi-Fi, and Internet protocols, such as Hypertext Transfer Protocol and Transmission Control Protocol. One example is the monitoring of the suitability of Fitbit,

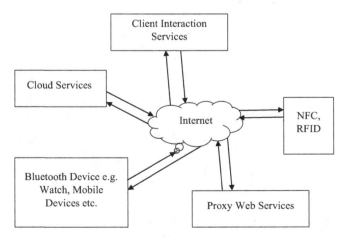

FIGURE 1.3   Method How Smart Objects Interact through IoT.

which loads the user's action count by his phone via 4G network to the user's account in the cloud. With proxy interaction, mobile users who happen to be close to an IoT-enabled device can view relevant information published by people interested in the Web service using their smartphone, just as they would when searching the Web [Roy Want et al. IEEE 2015].

RFID tool is a chief development in the field of embedded communication prototype, which permits policy of microchips for wireless local area network data communication. They support the automatic detection of whatever they are involved in acting as an electronic barcode. The passive RFID tags are not battery powered, and they use the power of the reader's debriefing signal to communicate the ID to the RFID reader.

IoT works on the root of Machine-to-Machine (M2M) infrastructures but is not restricted to it. M2M states communication between two technologies, deprived of human involvement. In IoT, even non-connected entities can become part of IoT, with a data communicating device, similar to a barcode or an RFID tag, detected through a device (might even be a smartphone detecting it), which ultimately is associated with the Internet. In IoT, non-intelligent objects, known as "things" in IoT terminology, become the collaborating nodes.

1.3 AI AND IoT

There is a clear link between IoT and AI. IoT is about connecting devices and using data generated from those devices. AI is about imitating intelligent behaviour on machines of all kinds. Apparently, a splash. Since IoT devices will generate a lot of data, AI will need to be operational to deal with this in large scale if we are to have the opportunity to understand the data. Data is only useful when creating action. For data to work, it needs to be added to content and art. IoT and AI together are in this context, namely, "connected intelligence" and not just connected devices.

Conventional methods of examining the coordinated data and generating action are not designed specifically to effectively process the enormous amounts of real-time information flow from the IoT devices. This is a place where AI-based evaluation and reply develop important for obtaining optimal value from that data.

The goal of applying AI to the IoT systems is to successfully implement an additional layer of intelligence across the IoT stack – from infrastructure to operating systems. AI is used, primarily, to create moderate or cooling sensors when the IoT network or each sensor fails, pinpoints them, or even creates new types of "visual sensors", or by using a Magnetic Resonance Imaging implant, a non-invasive imaging technique used, for example, to detect cancer cells and reproduce them as three-dimensional images.

The integration of these advanced technologies is predicted to drastically change the competitive landscape by requiring all companies to turn their product portfolios into reality. This will lead to the emergence of new ideas in all industries. The four most important changes that will be caused by the competitive environment are as follows:

- More Profits: The combined effect of AI and IoT will be of great benefit to many industries in terms of revenue and huge returns. IoT gadget producers, IoT data providers, and enterprise service providers based on smart sensors are expected to be on the winning end.
- Enhanced Safety Requirements: Real-time monitoring can help keep a tight check and thus prevent all kinds of failures or disasters. This will increase safety and security standards in general and increase efficiency. This will also help in reducing the loss of life and property damage.
- Decreased Costs: Rental equipment with smart sensors, sensor equipped with home appliances, smart electric meters, etc. It will lead to a reduction in the operating costs of families and businesses.
- Increased Customer Experience: Smart sensors come with a host of opportunities to improve customer experience. These sensors can read user preferences and adjust their values accordingly. For example, smart house thermostats can adapt to the ideal temperatures of various users living in the same house.

1.4 COGNITIVE IoT AND COVID-19 PANDEMIC

The world as a whole is battling with novel coronavirus; medical care officials are currently working vigorously to deliver the best service and all the essential amenities to help prevent the citizens from becoming infected and to rescue the people who are already contaminated. In such a situation, technological innovation is increasingly becoming a difference to healthcare systems. Nations are assembling all resources available to them and deploying cutting-edge technology to mitigate the consequences of COVID-19 as well as the profile of people at risk.

IoT and other technologies such as cloud and AI are extremely helpful in times of disaster. According to a study by researchers at Massachusetts Institute of Technology, including a geographic information system on IoT mobile data can assist epidemiologists in their search for patient zero and can help identify all people who have encountered infected patients. Technology can also help in identifying high-risk patients, which is why it can be a source of information for healthcare professionals to take the appropriate action.

Wearable technology has developed at a striking pace in the past few years and firmly established itself as an evolving product category. Since the COVID-19 cases continue growing by the day, healthcare stakeholders are discovering new tools and medications to help stem the tide. Changes that have taken place in technology and healthcare devices now are coming into their own and helping us fight against the pandemic. In such a scenario, wearable tech is playing a crucial role, which is effective at monitoring a number of physical and biological considerations such as cardiac rhythm, body temperature, artery pressure, movement, sleep, etc.

1.5 GLOBAL APPLICATIONS OF COGNITIVE IoT

IoT is an ecosystem and sensory system connected but a step forward in the technological development of the modern era is the introduction of cognition into IoT. Cognition means thinking that is a combination of understanding, learning, and thinking. Awareness is closely related to the functioning of the human brain, just as awareness is made by performing such tasks on the computer and in devices/sensors. In other words, cognitive IoT is the AI and devices connected to the calculated combination [8–10].

Examples:

- Food Industry: The food storage sector is an important area that requires access to real-time data and predictive analytics methods for maintaining food health; cognitive IoT is the best solution that provides such facilities on a platter. The temperature and environment of food storage units are continuously monitored with the help of sensors; further all the data is transferred to a cloud through the streaming analytics method.
- Garment Industry: In the clothing industry, IoT is extremely useful, such as administrators can gain access to such analytics from any

place on the production line as well as from any device. Additionally, this intelligent solution is a direct advantage to employees as well. Administrators can directly offer an incentive to them for accomplishing their daily production goals as their work is precisely catalogued, delivering them with immediate feedback. Additionally, there is a minimal investment in the infrastructure for factory companies.

• Health Industry: Connected health wearables are allowing the creation of hospitals "without barriers", wherein long-term medical treatment can be carried remotely by healthcare authorities to patients in their homes; therefore with the use of AI and IoT beds will be assigned to the patients who need more intensive care.

• Agriculture: The IoT is a network of devices with electronic devices, software, sensors, and connections that allow these objects to connect, interact, and exchange data. In this example, the IoT system does not need to be connected to the Internet. The benefits that farmers get from adapting the IoT system are twofold. It has helped farmers to reduce their costs and increase crop yields.

• Body Sensors: Smart sensors are very helpful in finding various bodily functions to maintain proper health. Many medical companies invest in therapeutic nerves that can assist patients in tracking their activities to improve their health; for example, these sensors can help monitor blood sugar levels and then release the insulin in case of emergency response.

1.6 CONCLUSION

As our lifestyle changes day by day, everyone wants to be free in their lives, and everyone wants a machine that works like a human being. Also, the world's population is growing, and resources are scarce, and the use of resources is very important too. Let's take an example. Now drinking water is a big problem in Asia due to water pollution and industrial pollution, and it is all based on a human dependence system. If we use an IoT tool in water management, then we can analyse water use, wastewater, and monitor water pollution. Also, there are some open-ended challenges for IoT comprehension such as:

• Energy-efficient sensing
• Security of data
• Secure reprogrammable networks and privacy
• Protocol for sensor networks

REFERENCES

1. Y. Yin, Y. Zeng, X. Chen, and Y. Fan, "The Internet of Things in Healthcare: An Overview" *Journal of Industrial Information Integration*, 1: 3–13, 2016.
2. M. Chen, Y. Zhang, M. Qiu, N. Guizani, and Y. Hao, "SPHA: Smart Personal Health Advisor Based on Deep Analytics" *IEEE Communications*, 56(3): 164–169, 2018.
3. L. Hu, et al., "Internet of Things Cloud: Architecture and Implementation" *IEEE Communications Magazine*, 54(12-Supp): 32–39, 2016.
4. M. Chen, et al., "Edge-CoCaCo: Toward Joint Optimization of Computation, Caching, and Communication on Edge Cloud" *IEEE Wireless Communications*, 25(3): 21–27, 2018.
5. G. Muhammad, S. K. M. M. Rahman, A. Alelaiwi, and A. Alamri, "Smart Health Solution Integrating IoT and Cloud: A Case Study of Voice Pathology Monitoring" *IEEE Communications Magazine*, 55(1): 69–73, 2017.
6. M. S. Hossain, "Patient State Recognition System for Healthcare Using Speech and Facial Expressions" *Journal of Medical Systems*, 40(12): 272:1–272:8, 2016.
7. M. S. Hossain, and G. Muhammad, Audio-visual Emotion Recognition Using Multi-directional Regression and Ridgelet Transform" *Journal on Multimodal User Interfaces*, 10(4): 325–333, 2016.
8. M. Chen, F. Herrera, and K. Hwang, "Cognitive Computing: Architecture, Technologies and Intelligent Applications" *IEEE Access*, 6: 19774–19783, 2018.
9. Cognitive IoT for Healthcare, https://www.ibmbigdatahub.com/blog/what-cognitive-iot
10. What Is Cognitive IoT, www.ibmbigdatahub.com

Cognitive Internet of Things

2

Smart Student Evaluation

2.1 EDUCATION AND INTERNET OF THINGS

Education organizations commonly experience the ill effects of restricted subsidizing, work issues, and poor thoughtfulness regarding real training. They, not at all like different associations, generally need or keep away from examination because of their subsidizing issues and the conviction that investigations do not have any significant bearing on their industry [1–3].

Internet of Things (IoT) not only delivers precious understanding but also democratizes such information across low-cost, low-power small appliances, which nevertheless provide high performance. Such technology supports in managing costs, enhancing the quality of learning, professional growth, and facility management development across rich assessments of crucial fields:

- Undergraduate response, accomplishment, and comportment
- Teacher response, performance, and conduct
- Facility supervising

The data informs them of ineffective strategies and practices, whether academic efforts or institutional merits. Removing these roadblocks makes them

DOI: 10.1201/9781003310341-2

more efficient. The information provided by IoT empowers teachers to deliver improved education. They have a window into the success of their strategy, the vision of their students, and other aspects of their work. IoT frees them from administrative tasks, so they can focus on their mission. It is self-crafted and secretarial and assists supervision by using features such as program flags or controls to ensure students remain loyal [4,5].

2.1.1 IoT and Education Institution

IoT stipulates mentors with laid-back access to influential educational tools. Educators may use IoT to customize on-demand course material for students. IoT advances the professional development of mentors since they accurately see what works and learn to formulate improved tactics, instead of merely repeating old or unproductive tactics.

Furthermore, IoT boosts the knowledge base which may be used to formulate academic standards and practices. IoT in the field of academic research agonizes from accuracy issues and a general lack of data.

IoT sensors wrinkle the data; communication components spread the information gathered. In the future, vast numbers of sensors will be installed which will harvest massive amounts of data and encompass concealed information (knowledge), which will help to devise a better system. Machine learning (ML) and data mining techniques play a crucial role in extracting useful information from gathered data.

Earlier, a collection of hardware, software, and online services came into the market, which claimed reforms to classrooms and tutoring methods. But the true commotion of education is yet to attain.

The institution aims to develop a platform that provides real-time reviews and assistance in such a way that online tutors be better at tutoring. As an example, the system would figure out if a student's answer to the notions is following a pattern of misinterpretation. By providing the premature caveat to teachers, the platform could help exclude the issues that are further into teaching-learning [6].

"If we can aim to shape the performance of the teacher — the teacher being the significant input into a child's learning — then you're creating something truly powerful", says Tom Hooper, founder of Third Space Learning.

"With the increasing capabilities of machine learning, there is a unique opportunity to personalize learning to individual students", says Erik Choi, Principal Researcher at Brainly.

With the cumulative competencies of ML, there is an exceptional chance to personalize learning for every student. Even we can predict the performance of the student so he/she can know in which section they need improvement and can perform well.

Each student can access information that would help them along their own one-of-a-kind manner of its expectation to study and adapt. Consequently student can choose own learning path as per their learning pace. Figure 2.1 depicts how the data collected from IoT apparatus goes through preprocessing (Dimension Reduction) and Statistical Modelling, that is, ML, and the result has been helping in decision-making for the organization.

Reading material, student books, and educational material are usually custom-made for the students and printed in large numbers [7]. In each case, not the whole faculty and schools have a similar instructing style. By applying Statistical Modelling, teachers and schools will be able to create textbooks and exercises that are made to order to the needs of their specific courses and students.

IoT simplifies the customization of tutoring to provide access to what students need. The student merely utilizes the system, and student performance data principally outlines the education system design [8]. This thing, collectively with organization and educator, that is, optimized delivery of highly effective education while reducing costs.

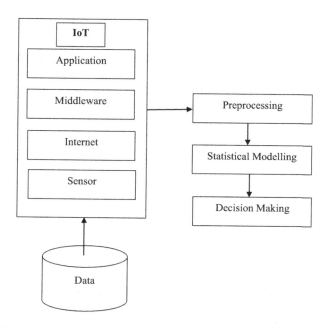

FIGURE 2.1   IoT and Statistical Modelling.

2.2 MACHINE LEARNING CLASSIFIERS FOR SMART EDUCATION

Today's roles of analytics on serving personalities to make intelligence of the learning measures and smart environments on providing feedback and diverse "smart" functionalities have pinched the interest of many scholars and practitioners. Most recent improvements in these two zones set their convergence as an exceptionally encouraging exploration region [9,10].

After going through several pieces of literature, we came across the bottlenecks of the existing tutoring system as follows:

- A prominent issue with any development programme is their monitoring, cost involved in the programme, and manpower availability.
- Better implementation of student development programmes can be ensured only if those responsible for actual implementation are paid reasonably well, appropriately trained, and sufficiently motivated. But this has not yet been done.
- Value of mean squared error (MSE) is high while modelling between impendent and dependent parameters; hence, accuracy decreases the scope of smart tutoring.
- Connected devices can help make life easier for students with special needs. For instance, a visually impaired student who is given a special card that, when registered by a computer, automatically enlarges the font size. Rather than having to call a teacher over for help – costing both the student and the teacher time they could be using more productively – the student can take care of the issue, which also builds self-confidence and promotes independence.

2.2.1 ML Classifiers

The institution's goal is to develop a platform that provides real-time feedback and assists online tutors to become better at tutoring. For example, the system will perceive if a student's response to a concept follows a pattern of misinterpretation [11,12]. By giving premature warnings to teachers, the platform can help preclude problems further in the teaching-learning process. We have proposed smart tutoring based on regression Statistical Modelling; the layout of our proposed tutoring scheme is as follows:

Figure 2.2 shows the layout of the proposed scheme. IoT devices will gather the student performance index, that is, their class test marks, and gathered data

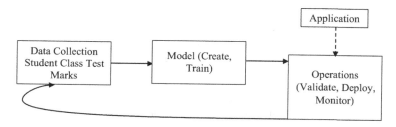

FIGURE 2.2 Proposed Workflow.

will be stored in the storage device. Upon gathering data (knowledge base), we will apply our proposed algorithm, which will train the data and predict the student performance.

Predicted student performance provides positive feedback to the administrator of the institute so that they can identify slow learner and fast learner students, and, accordingly, they can do some assignments to improve the performance of slow learner students.

Curve fitting, additionally, acknowledged as regression analysis, is used to discover the "best fit" line-up or curve for a series of data points. Much of the time, the curve fit is going to produce an equation that can be used to locate points anywhere along the curve. In certain cases, you might not be concerned about finding an equation. However, you may simply wish to use the curve fit to smooth the data and enhance the appearance of your plot. MATLAB tool delivers curve fits that can be used in both of the scenarios above (Figure 2.3).

Proposed Algorithm

Step-1. Input training dataset.
Step-2. Input test dataset.
Step-3. Process training dataset to find fitness function between dependent and independent variables of the training dataset.
Step-4. Apply polynomial curve fitting.
Step-5. Find optimal fitness function as MSE will be lesser.
Step-6. Apply regression Statistical Modelling as per optimal fitness function.
Step-7. Predicted data as output.
Step-8. Apply classifier to predicted output.

Multiple Regression (K-Nearest Neighbor (KNN))

Regression (featTrain, classTrain, featTest, classTest, featName, classifier)
  /*featTrain – A NUMERIC matrix of training features (N × M)

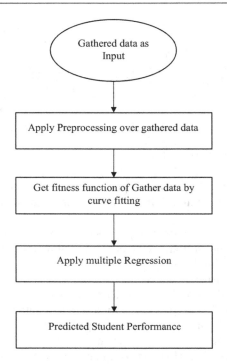

FIGURE 2.3   Proposed System Flow.

classTrain – A NUMERIC vector representing the values of the dependent variable of the training data (N × 1)
   featTest – A NUMERIC matrix of testing features (Nts × M)
   classTest – A NUMERIC vector representing the values of the dependent variable of the testing data (Nts × 1)
   featName – The CELL vector of string representing the label of each features (1 × M) cell*/
   //Classification algorithm as KNN Regression
   Step-1 NNBestFeat = floor(Datapoints()/10) //nearest neighbour
   Step-2 trainModel = KNN Regression model
   Step-3 NNSearch = Initialize search function for KNNReg as linear search
   //Set the distance measure for NNSearch
   Step-4 distFunc = Euclidean distance (or similarity) function
   Step-5 trainModel.setNearestNeighbourSearchAlgorithm (NNSearch)
   Step-6 trainModel.setKNN(NNBestFeat)

Curve fitting is considered to be the procedure of constructing a curve, or a mathematical function that contains the best fit to a series of data points, conceivably subject to constraints. Curve fitting can comprise either interpolation, where a strict fit to the data is essential, or smoothing, in which a "smooth" function is constructed that just about fits the data.

Regression analysis focuses mainly on statistical trends questions such as how much inconsistency exists in the right curve for data detected by random errors. The included curves can be used as a data visual aid, to enter activity values where no data is available, and to integrate the relationship between two or more variables. Extrapolation refers to the use of a curved curve that exceeds the width of the visual data and is less questionable because it can reflect the method used to construct the curve as it reflects the data.

2.3 IMPLEMENTATION USING MATLAB TOOL

2.3.1 Dataset Used and Curve Fitting Tool

For the implementation of smart evaluation, MATLAB 2019b has been used; the dataset used in our experiment is data gathered from an engineering college student dataset. A snippet of the dataset is as follows (Figure 2.4):

Figure 2.5 shows the MATLAB curve fitting tool, which generates the different parameters of the fitness model as we have discussed in the solution methodology section. We have highlighted the parameters in the figure too.

A	B	C	D
RollNo	CT1	CT2	FINAL
NUMBER ▾	NUMBER ▾	NUMBER ▾	NUMBER ▾
3422214001	81.66666667	83.33333333	1
3422214002	80.83333333	80.83333333	1
3422214003	85.83333333	81.66666667	1
3422214004	79.16666667	72.5	1
3422214006	81.66666667	70.83333333	1
3422214007	80	81.66666667	1
3422214008	70.83333333	74.16666667	1

FIGURE 2.4 Snippet of Gathered Data.

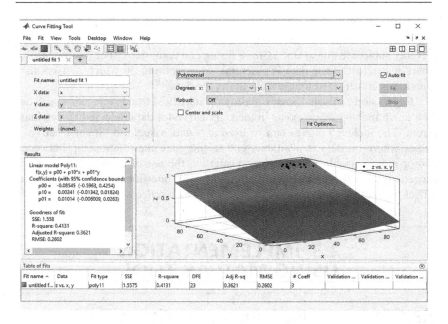

FIGURE 2.5 Curve Fitting Applied on Input Data.

2.3.2 Classification Learner Tool

Figure 2.6 shows the MATLAB classification learner output as the output of the proposed system passed through Support Vector Machine (SVM) Classifier, meant to check the accuracy of predicted values, as it gives output with 88.5% accuracy.

Figure 2.7 shows the performance plot, from there we can say that as the training dataset size increases, the accuracy of the proposed algorithm also increases.

2.4 SUMMARY

In IoT, sensors gather data, and communication components relay the information gathered. In future, enormous numbers of sensors will be deployed which will produce huge amounts of data and contain hidden information (knowledge), which can help to make a better system. In this case, ML and

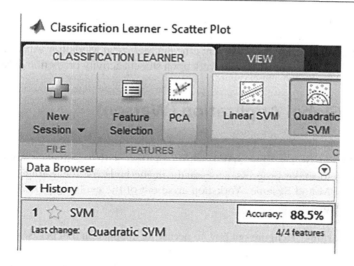

FIGURE 2.6   Accuracy of SVM Classifier.

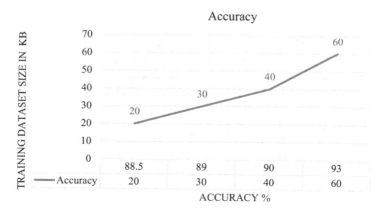

FIGURE 2.7   Performance Plot.

data mining technique will play an important role to extract knowledge from captured data from an IoT system.

The institution's goal is to develop a platform that provides real-time feedback and assistance to online tutors. If the institution has predicted data, then they can categorize the student as Good, Average, Best, etc. and apply remedies that can improve the student performance. Devices used in IoT produce a huge amount of data which attracts attention towards applying Statistical

Modelling over data produced. Classifier output shows that our proposed algorithm prediction accuracy is 88.5%.

In future, we can apply different efficient Statistical Modelling Approach to increase the accuracy, and we can work upon student activity prediction in future.

2.4.1 Global Application (Watson)

Watson's objective is to bring adaptive learning to the next level, with the establishment of teaching and learning platforms that influence cognitive computing to make progress necessarily in methods for the young children to learn. IBM and Sesame Workshop arise out of the evolving possible tools; some of them are:

1. Learning: To figure out how to peruse an application that peruses with the kid, making intuitive game encounters and utilizing their own words. For instance, the kid is told, "We should take care of business with the canine". Watson then, at that point, examines the youngster's reaction continuously, and it progressively adjusts to their inclinations dependent on content so that perusing is more enjoyable by finishing stories with characters, creatures, and toys that the kid likes.
2. Toy: It could be the anointed Elmo himself, who can listen and respond, using the child's details to create recreational activities. With the power of Watson's cognitive computer embedded in Elmo, the toy adapts to the child's developmental skills over time (e.g., making letters or counting up to 20). Once these skills are learned, Elmo can provide new learning activities for further learning.
3. Smart Classroom: A request which meets the recent discoveries made of science learning in order to assist educators design learning lessons for both business and leisure travel; individual requirements of every student might be developed. Watson concentrates on educational objectives which are most suitable for a specific teacher as well as to the child at the time. The tool can also proactively approach pinpoint concepts that some students may need, and it recommends different learning experiences as well as speeding up learning strategies.

REFERENCES

1. Chun-Wei Tsai, Chin-Feng Lai, Ming-Chao Chiang, et al., *Data Mining for Internet of Things: A Survey*. IEEE, 2014.

2. Najah Abu Ali, et al., *Data Management for The Internet of Things: Green Directions*. IEEE, 2012.
3. J. Baliga, R. W. A. Ayre, K. Hinton, and R. S. Tucker, "Green Cloud Computing: Balancing Energy in Processing, Storage, and Transport" *Proceedings of the IEEE*, 99(1): 149–167, January 2011.
4. Feng Wang, Liang Hu, Jin Zhou, and K. Zhao, "A Survey from the Perspective of Evolutionary Process in the Internet of Things Hindawi Publishing Corporation" *International Journal of Distributed Sensor Networks*, 11(3): 462752, 2015.
5. Shen Bin, Liu Yuan, et al., *Research on Data Mining Models for the Internet of Things*. IEEE, 2010.
6. Jayavardhana Gubbi, Rajkumar Buyya, et al., *Internet of Things (IoT): A Vision, Architectural Elements, and Future Directions*. Elsevier, 2013.
7. Edmund W. Schuster, Sumeet Kumar, et al., *Infrastructure for Data-Driven Agriculture: Identifying Management Zones for Cotton using Statistical Modeling and Machine Learning Techniques*. IEEE, 2011.
8. Michail N. Giannakos, Demetrios G. Sampson, and Łukasz Kidziński, *Introduction to Smart Learning Analytics: Foundations and Developments in Video-based Learning*. Springer, 2016.
9. J. Gehrke and S. Madden, "Query Processing in Sensor Networks" *IEEE Pervasive Computing*, 3(1): 46–55, March 2004.
10. L. Chen, M. Tseng, and X. Lian, "Development of Foundation Models for Internet of Things" *Frontiers of Computer Science in China*, 4(3): 376–385, September 2010.
11. R. Cattell, "Scalable SQL and NoSQL Data Stores" *SIGMOD Record*, 39(4): 12–27, December 2010.
12. F. Chang et al., "Bigtable: A Distributed Storage System for Structured Data" *ACM Transactions on Computer Systems (TOCS)*, 26(2): 1–26, June 2008.

Cognitive Internet of Things

3

Chronic Disease Prediction

3.1 CHRONIC DISEASE AND HUMAN HEALTH

The healthcare sector is currently being transformed by way of the capability to record huge amounts of information regarding sick people, and the vast quantity of information contained in human beings collected makes it impossible for investigation [1]. Machine learning offers a way to discover motifs and reasons about data, which allows medical professionals to relocate to customized medical treatment, commonly referred to as precision medicine. There are many opportunities for exactly how machine learning could be used in health care, and everything of them must rely on having adequate data and authorization to use it. Data mining can be regarded as a superset of several different approaches to obtain insight from data. This could consist of conventional statistical methods and machine learning [2,3].

The data produced by the healthcare organizations is enormous and complicated, due to which it is hard to analyse the data to make a significant decision concerning patient health. This data includes details regarding hospitals, patients, medical claims, medical care costs, etc. So, it is necessary to produce an effective tool for analysing and how to extract vital information from

DOI: 10.1201/9781003310341-3

this complex data [4,5]. Such an analysis of health information may enhance health care. The result of data mining technologies is to provide advantages for the medical treatment organization for a way to group the patients experiencing the comparable type of illnesses or health issues to ensure that healthcare organizations deliver them the most efficient treatments. It might also be useful for forecasting the length of a patient's stay in the hospital, medical diagnosis [6,7], and planning for efficient information system management. The latest technologies that are available are being used in the medical sector to improve healthcare services cost-effectively. Data mining techniques are also utilized to analyse the different factors which are responsible for the diseases: for example, type of food, various working environments, educational attainment, living conditions, the accessibility of the clean water, medical care services, and intellectual, ecological, and farming factors. Numerous mobile apps are being used to improve health and provide diet plans. Mobile apps, such as HealthifyMe, Google Fit, etc., utilize a machine learning algorithm, employed over the individual behaviour data, which accordingly trains customized (user-specific) health plan [8].

3.1.1 Chronic Disease Monitoring

Heart rate (HR) is one of the foremost symptoms that physicians regularly detect for heart-related disorders, such as different types of arrhythmias. HR and HR variants (HRV) are typically extracted from the electrocardiogram. Internet of Things (IoT) devices gather data and apply machine learning algorithms[9–14] to extract knowledge from gathered data as depicted in Figure 3.1.

There is a different performance metric to evaluate the efficacy of machine learning classifiers as follows:

The performance measures chosen to assess the classification techniques are precision and recall. Precision (P) has been defined as the number of true positives (Tp) [15] over the number of true positives and the number of false positives (Fp):

$P = Tp/(Tp+Fp)$ ....(1)

Recall (R) has been defined as the number of true positives (Tp) over the number of true positives and the number of false negatives (Fn):

$R = Tp/(Tp+Fn)$ ....(2)

Accuracy is also intended for comparing the classification results, which is defined as (see Figure 3.2):

$Accuracy = (Tp+Tn)/(Tp+Tn+Fp+Fn)$ ....(3)

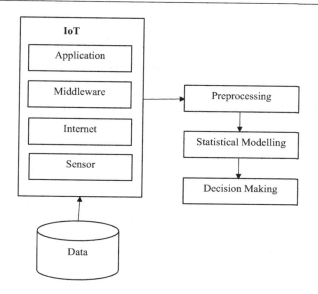

FIGURE 3.1   IoT and Statistical Modelling.

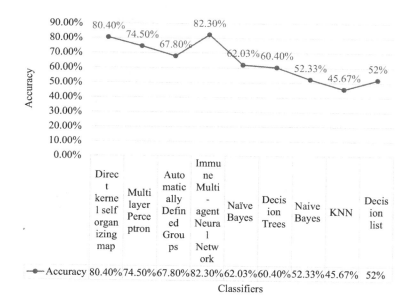

FIGURE 3.2   Comparison of Different Classifiers Used in Literature.

3.2 DISEASE PREDICTION AND MACHINE LEARNING

3.2.1 Bottlenecks and Prediction Model

- Difficulties in the field of medicinal finding are a gigantic quantity of data produced from medical centres, so there is a certain amount of noise incorporated in data collected, which affects the machine learning classification algorithm performance.
- Getting into a deep-rooted drive, applying machine learning entails four basic steps: (1) data preprocessing, (2) feature selection, (3) selecting the correct machine learning classifier, and (4) validation of predicted values. While we know data preprocessing is the key step, this procedure includes the elimination of noise present within the dataset and recognition of missing data values (Figure 3.3).

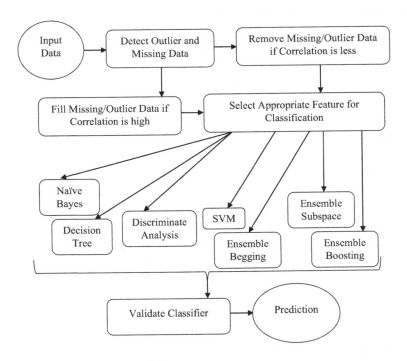

FIGURE 3.3 Prediction Model.

3.2.2 Naive Bayes Machine Learning Classifier

Naive Bayes (NB): NB classifiers are exceedingly scalable, needing numerous parameters linear in the number of variables (features/predictors) in a learning problem. NB is a simple technique for constructing classifiers: models that assign class labels to problem instances, represented as vectors of feature values, where the class labels are drawn from some finite set. It is not a single algorithm for training such classifiers but a family of algorithms based on a common principle: all NB classifiers assume that the value of a particular feature is independent of the value of any other feature, given the class variable. For the mathematical understanding of the NB classifier, we must know the following terms:

- Conditional Probability: a measure of the probability of event A occurring given that another event has occurred. For example, "what is the probability that it will rain given that it is cloudy?" is an example of conditional probability.
- Joint Probability: a measure that calculates the likelihood of two or more events occurring at the same time.
- Proportionality: refers to the relationship between two quantities that are multiplicatively connected to a constant, or in simpler terms, whether their ratio yields a constant.
- Bayes Theorem: describes the probability of an event (posterior) based on the prior knowledge of conditions that might be related to the event.

The NB classifier is inspired by Bayes Theorem which states the following equation:

$$P(A \mid B) = \frac{P(B \mid A) * P(A)}{P(B)} \tag{1}$$

This equation-1 can be rewritten using X (input variables) and y (output variable) to make it easier to understand. In plain English, this equation is solving for the probability of y given input features X.

$$P(y \mid X) = \frac{P(X \mid y) * P(y)}{P(X)} \tag{2}$$

Because of the naive assumption that variables are independent given the class, we can rewrite $P(X|y)$ as follows:

$$P(X \mid y) = P(x_1 \mid y) * P(x_2 \mid y) * \ldots * P(x_n \mid y) \tag{3}$$

Since we are solving for y, $P(X)$ is a constant which means that we can remove it from the equation and introduce a proportionality. This leads us to the following equation:

$$P(y \mid X) \propto P(y) * \prod_{i=1}^{n} P(x_i \mid y) \tag{4}$$

The goal of NB is to choose the class y with the maximum probability:

$$y = \operatorname{argmax}_y \left[P(y) * \prod_{i=1}^{n} P(x_i \mid y) \right] \tag{5}$$

Argmax is simply an operation that finds the argument that gives the maximum value from a target function. In this case, we want to find the maximum y value.

3.3 HEART DISEASE PREDICTION USING MATLAB TOOL

3.3.1 Dataset and Source

For the experimental evaluation, we took data from the subsequent repository: https://archive.ics.uci.edu/ml/datasets/Heart+Disease.

Figure 3.4 shows the dataset snippet and dependent and independent features. The description of the features can be obtained from the given URL.

age	sex	cp	trestbps	chol	fbs	restecg	thalach	exang	oldpeak	slope	ca	thal	num
63	1	1	145	233	1	2	150	0	2.3	3	0	6	absent
53	1	4	140	203	1	2	155	1	3.1	3	0	7	present
56	1	3	130	256	1	2	142	1	0.6	2	1	6	present
52	1	3	172	199	1	0	162	0	0.5	1	0	7	absent
58	0	1	150	283	1	2	162	0	1	1	0	3	absent
60	1	4	117	230	1	0	160	1	1.4	1	2	7	present
61	1	3	150	243	1	0	137	1	1	2	0	3	absent
59	1	3	150	212	1	0	157	0	1.6	1	0	3	absent

Independent
Variable

Dependent Variable

FIGURE 3.4   Dataset Snippet.

3.3.2 Classifiers and Accuracy

Classifier: Kernel Naïve Bayes

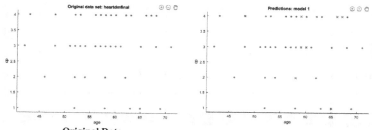

Original Data

Predicted Data

| 1 ☆ Naive Bayes | Accuracy: **84.7%** |
| Last change: Kernel Naive Bayes | 13/13 features |

Classifier: Linear SVM

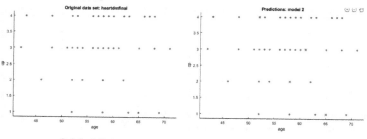

Original Data

Predicted Data

| 2 ☆ SVM | Accuracy: **90.6%** |
| Last change: Linear SVM | 13/13 features |

Quadratic SVM

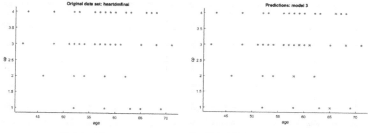

Original Data

Predicted Data

| 3 ☆ SVM | Accuracy: 85.9% |
| Last change: Quadratic SVM | 13/13 features |

Decision Tre

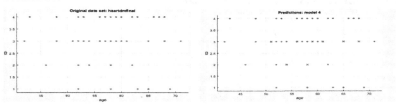

Original Data   **Predicted Data**

4 ☆ Tree   Accuracy: 84.7%
Last change: Fine Tree   13/13 features

KNN

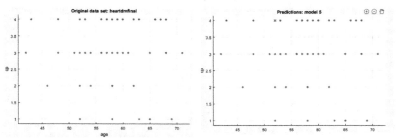

Original Data   **Predicted Data**

5 ☆ KNN   Accuracy: 98.8%
Last change: Fine KNN   13/13 features

Ensemble (Boosting)

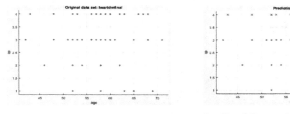

Original Data   **Predicted Data**

6 ☆ Ensemble   Accuracy: 54.1%
Last change: Boosted Trees   13/13 features

Ensemble (Subspace)

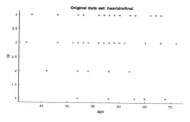 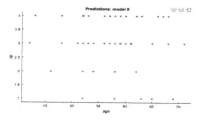

Original Data   **Predicted Data**

8 ☆ Ensemble   Accuracy: 89.4%
Last change: Subspace Discrimin...   13/13 features

Ensemble Begging

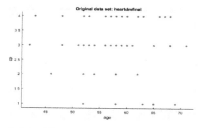 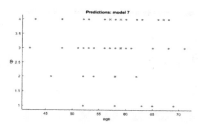

Original Data   **Predicted Data**

7 ☆ Ensemble   Accuracy: 90.6%
Last change: Bagged Trees   13/13 features

Matlab Code:
Code Fragment for Ensemble Classifier:

```
%Ensemble Classifier
clc
clear all
data = readtable('train.csv');
%%-------------Building Classifier--------------------------
%--------------------------Code--------------------------
classification_model = fitcensemble(data,'status');

cv = cvpartition(classification_model.NumObservations,
'HoldOut', 0.4);
cross_validated_model =
crossval(classification_model,'cvpartition',cv);
```

```
Predictions = predict(cross_validated_model.
Trained{1},data(test(cv),1:end-1))
%
%%%testset to print test set data uncomment below line
code
%data(test(cv),1:end-1)

%%--------------Analyzingthepredictions--------------------
%-------------------------- Code --------------------------
confusionmatval =
confusionmat(cross_validated_model.Y(test
(cv)),Predictions);
% ---------Calculate Accuracy --------------------------
TP=confusionmatval(1,1);
FN=confusionmatval(1,2);
FP=confusionmatval(2,1);
TN=confusionmatval(2,2);
Accuracy=((TP+TN)/(TP+TN+FP+FN))
```

Matlab Code for Decision Tree Classifier:

```
clear all

%%---------------Importing the dataset----------------------
%------------------------- Code --------------------------
data = readtable('dataset/heart dm final.csv');

classification_model = fitctree(data,'num');

%%--------------Test and Trainsets--------------------------
%------------------------- Code --------------------------
cv = cvpartition(classification_model.NumObservations,
'HoldOut', 0.4);
cross_validated_model =
crossval(classification_model,'cvpartition',cv);

%%-------------Making Predictions for Test sets--------------
%------------------------- Code --------------------------
Predictions = predict(cross_validated_model.
Trained{1},data(test(cv),1:end-1));
%% tree analysis
view(classification_model,'mode','graph')
```

Matlab Code for Preprocessing of Input Data:
```
clear all
```

```
%%---------------Importing the dataset----------------------
%----------------------------Code--------------------------
data = readtable('processed_cleveland_data.csv');
%%--------------Data Preprocessing---------------------------
%%-------------Feature Scaling-------------------------------
%--------------Method1:Standardization---------------------
%-----------------------Code-------------------------------
stand_var1 = (data.Var1 - mean(data.Var1))/std(data.
Var1);
data.Var1 = stand_var1;
stand_var2 = (data.Var2 - mean(data.Var2))/std(data.Var2);
data.Var2 = stand_var2;
%%---------------Dimensionality Reduction--------------------
%%-----------------------PCA--------------------------------
%-----------------------Code-------------------------------
class_labels = data.Var14;
data = table2array(data(:,1:end-1));
[coeff,score,latent,tsquared,explained,mu] = pca(data);
Var1 = score(:,1);
Var2 = score(:,2);
data = table(Var1, Var2, class_labels);
```

Logistic Regression:

```
function [trainedClassifier, validationAccuracy] =
trainClassifier(trainingData)
% [trainedClassifier, validationAccuracy] =
trainClassifier(trainingData)
% Input:
% trainingData: a table containing the same predictor and
response
% columns as imported into the app.
% Output:
% trained classifier: a struct containing the trained
classifier. The
% struct contains various fields with information about
the trained
% classifier.
%
%trainedClassifier.predictFcn: a function to make
predictions on new data.
% Extract predictors and response
% This code processes the data into the right shape for
training the
% model.
inputTable = trainingData;
```

```matlab
predictorNames = {'age', 'sex', 'cp', 'trestbps', 'chol',
'fbs', 'restecg', 'thalach', 'exang', 'oldpeak', 'slope',
'ca', 'thal'};
predictors = inputTable(:, predictorNames);
response = inputTable.num;
isCategoricalPredictor = [false, false, false, false, false,
false, false, false, false, false, false, false, false];
% Train a classifier
% This code specifies all the classifier options and
trains the classifier.
% For logistic regression, the response values must be
converted to zeros
% and ones because the responses are assumed to follow a
binomial
% distribution.
% 1 or true = 'successful' class
% 0 or false = 'failure' class
% NaN - missing response.
% Compute the majority response class. If there is a
NaN-prediction from
% fitglm, convert NaN to this majority class label.
numSuccess = sum(response == successClass);
numFailure = sum(response == failureClass);
if numSuccess > numFailure
  missingClass = successClass;
else
  missingClass = failureClass;
end
successFailureAndMissingClasses = [successClass;
failureClass; missingClass];
isMissing = isnan(response);
zeroOneResponse = double(ismember(response,
successClass));
zeroOneResponse(isMissing) = NaN;
% Prepare input arguments to fitglm.
concatenatedPredictorsAndResponse = [predictors,
table(zeroOneResponse)];
% Train using fitglm.
GeneralizedLinearModel = fitglm(...
  concatenatedPredictorsAndResponse, ...
'Distribution', 'binomial', ...
'link', 'logit');
% Convert predicted probabilities to predicted class
labels and scores.
convertSuccessProbsToPredictions = @(p)
successFailureAndMissingClasses( ~isnan(p).*( (p<0.5) + 1 )
+ isnan(p)*3 );
```

```
returnMultipleValuesFcn = @(varargin)
varargin{1:max(1,nargout)};
scoresFcn = @(p) [1-p, p];
predictionsAndScoresFcn = @(p) returnMultipleValuesFcn(
convertSuccessProbsToPredictions(p), scoresFcn(p) );

% Create the result struct with predict function
predictorExtractionFcn = @(t) t(:, predictorNames);
logisticRegressionPredictFcn = @(x)
predictionsAndScoresFcn( predict(GeneralizedLinearModel,
x) );
trainedClassifier.predictFcn = @(x)
logisticRegressionPredictFcn(predictorExtractionFcn(x));
% Add additional fields to the result struct
trainedClassifier.RequiredVariables = {'age', 'ca',
'chol', 'cp', 'exang', 'fbs', 'oldpeak', 'restecg',
'sex', 'slope', 'thal', 'thalach', 'trestbps'};
trainedClassifier.GeneralizedLinearModel =
GeneralizedLinearModel;
trainedClassifier.SuccessClass = successClass;
trainedClassifier.FailureClass = failureClass;
trainedClassifier.MissingClass = missingClass;
trainedClassifier.ClassNames = {successClass;
failureClass};
trainedClassifier.About = 'This struct is a trained model ';
% Extract predictors and response
% This code processes the data into the right shape for
training the
% model.
inputTable = trainingData;
predictorNames = {'age', 'sex', 'cp', 'trestbps', 'chol',
'fbs', 'restecg', 'thalach', 'exang', 'oldpeak', 'slope',
'ca', 'thal'};
predictors = inputTable(:, predictorNames);
response = inputTable.num;
isCategoricalPredictor = [false, false, false, false,
false, false, false, false, false, false, false, false,
false];

% Perform cross-validation
KFolds = 5;
cvp = cvpartition(response, 'KFold', KFolds);
% Initialize the predictions to the proper sizes
validationPredictions = response;
numObservations = size(predictors, 1);
numClasses = 2;
validationScores = NaN(numObservations, numClasses);
```

```
for fold = 1:KFolds
  trainingPredictors = predictors(cvp.training(fold), :);
  trainingResponse = response(cvp.training(fold), :);
  foldIsCategoricalPredictor = isCategoricalPredictor;

  % Train a classifier
  % This code specifies all the classifier options and
trains the classifier.
  % For logistic regression, the response values must be
converted to zeros
  % and ones because the responses are assumed to follow a
binomial
  % distribution.
  % 1 or true = 'successful' class
  % 0 or false = 'failure' class
  % NaN - missing response.
  successClass = double(1);
  failureClass = double(0);
  % Compute the majority response class. If there is a
NaN-prediction from
  % fitglm, convert NaN to this majority class label.
  numSuccess = sum(trainingResponse == successClass);
  numFailure = sum(trainingResponse == failureClass);
  if numSuccess > numFailure
    missingClass = successClass;
  else
    missingClass = failureClass;
  end
  successFailureAndMissingClasses = [successClass;
failureClass; missingClass];
  isMissing = isnan(trainingResponse);
  zeroOneResponse = double(ismember(trainingResponse,
successClass));
  zeroOneResponse(isMissing) = NaN;
  % Prepare input arguments to fitglm.
  concatenatedPredictorsAndResponse = [trainingPredictors,
table(zeroOneResponse)];
  % Train using fitglm.
  GeneralizedLinearModel = fitglm(...
    concatenatedPredictorsAndResponse, ...
    'Distribution', 'binomial', ...
    'link', 'logit');

  % Convert predicted probabilities to predicted class
labels and scores.
  convertSuccessProbsToPredictions = @(p) successFailure
AndMissingClasses( ~isnan(p).*( (p<0.5) + 1 ) + isnan(p)*3 );
```

```
returnMultipleValuesFcn = @(varargin)
varargin{1:max(1,nargout)};
  scoresFcn = @(p) [1-p, p];
  predictionsAndScoresFcn = @(p) returnMultipleValuesFcn(
convertSuccessProbsToPredictions(p), scoresFcn(p) );

  % Create the result struct with predict function
  logisticRegressionPredictFcn = @(x)
predictionsAndScoresFcn( predict(GeneralizedLinearModel,
x) );
  validationPredictFcn = @(x) logisticRegressionPredictFcn
(x);

  % Add additional fields to the result struct

  % Compute validation predictions
  validationPredictors = predictors(cvp.test(fold), :);
  [foldPredictions, foldScores] =
validationPredictFcn(validationPredictors);

  % Store predictions in the original order
  validationPredictions(cvp.test(fold), :) =
foldPredictions;
  validationScores(cvp.test(fold), :) = foldScores;
end

% Compute validation accuracy
correctPredictions = (validationPredictions == response);
isMissing = isnan(response);
correctPredictions = correctPredictions(~isMissing);
validationAccuracy =
sum(correctPredictions)/length(correctPredictions);
```

3.4 SUMMARY

Disease diagnosis is one of the applications where data mining tools provide successful results (Figure 3.5). Heart disease has been the leading cause of death all over the world in the past ten years. Several researchers are using statistical and data mining tools to help healthcare professionals in the diagnosis of heart diseases (Table 3.1). From different experiments, among different classifiers, KNN achieved 98.8% accuracy. Different machine learning classifiers can be merged to improve classification accuracy.

TABLE 3.1   Comparison

S. NO.	CLASSIFIER	ACCURACY (%)
1	NB	84.7
2	Ensemble (Begging)	54.1
3	Ensemble (Boosting)	90.6
4	Ensemble (Subspace)	89.4
5	Ensemble (KNN)	89.4
6	Fine KNN	98.8
7	Decision Tree	84.7
8	Linear SVM	90.6
9	Quadratic SVM	85.9
10	Weighted KNN	97.6
11	Cubic KNN	83.5
12	Cosine KNN	85.9
13	Logistic Regression	87.1

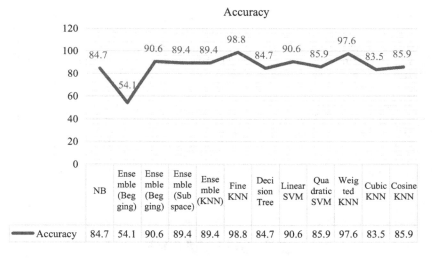

FIGURE 3.5   Accuracy of Different Machine Learning Classifier.

REFERENCES

1. Theodora S. Brisimi, Tingting Xu, Taiyao Wang, Wuyang Dai, y William G. Adams, and Ioannis Ch. "Paschalidis, Predicting Chronic Disease Hospitalizations from Electronic Health Records: An Interpretable Classification

Approach" *Proceedings of the IEEE*, 106(4), DOI: 10.1109/JPROC.2017.2789319, April 2018.

2. T. Rajasekaran, Priya R. Tharani, V. Pavithra, and P. Sudharshanam, "Heart Attack Prediction for Diabetic's Patient using Data Analytics" *South Asian Journal of Engineering and Technology* 2(17): 103–110, ISSN No: 2454–9614, 2016.

3. A. Upadhyay, and V. R. Patel, Comparative Study - Prediction of Diabetes and Heart Disease using Data Mining Approaches" *International Journal of Engineering Technology, Management and Applied Sciences*, 4(1), ISSN 2349–4476, 2016.

4. D. Asir Antony Gnana Singh, "Dimensionality Reduction using Genetic Algorithm for Improving Accuracy in Medical Diagnosis" 2016 MECS I.J. *Intelligent Systems and Applications*, 1: 67–73, 2016.

5. S. Perveena, Md. Shahbaza, A. Guergachib, and K. Keshavjee, "Performance Analysis of Data Mining Classification Techniques to Predict Diabetes, Symposium on Data Mining Applications" SDMA2016, 1877-0509 © 2016 Published by ELSEVIER B.V, 2016.

6. G. Parthiban, A. Rajesh, and S. K. Srivatsa, "Diagnosing Vulnerability of Diabetic Patients to Heart Diseases using Support Vector Machines" *International Journal of Computer Applications* (0975 – 888), 48(2): 45–49, June 2012.

7. Mai Shouman, Tim Turner, and Rob Stocker, Using Data Mining Techniques in Heart Disease Diagnosis and Treatment, 978-1-4673-0483-2 c 2012 IEEE.

8. P. R. Pawar, and D. Vora, "Prediction of Heart Disease for Diabetic Patients using Genetic Neural Network" *International Journal on Recent and Innovation Trends in Computing and Communication*, 3(7): 5028–5030, ISSN: 2321–8169, 2015.

9. Santhanam, T., & Padmavathi, M. S. (2015). Application of K-means and genetic algorithms for dimension reduction by integrating SVM for diabetes diagnosis. *Procedia Computer Science*, 47, 76–83.

10. A. Aljumah, and Md Siddiqui, "Data Mining Perspective: Prognosis of Life Style on Hypertension and Diabetes" *The International Arab Journal of Information Technology*, 13(1), 2015.

11. C. Kalaiselvi, and G. M. Nasira, "Prediction of Heart Diseases and Cancer in Diabetic Patients Using Data Mining Techniques" *Indian Journal of Science and Technology*, 8(14), DOI: 10.17485/ijst/2015/v8i14/72688, ISSN (Print): 0974–6846 ISSN (Online): 0974–5645, July 2015.

12. Radha, P. (2014). Diagnosing heart diseases for type 2 diabetic patients by cascading the data mining techniques. *International Journal on Recent and Innovation Trends in Computing and Communication*, 2(8), 2503–2509.

13. Gaganjot Kaur, and Amit Chhabra, "Improved J48 Classification Algorithm for the Prediction of Diabetes" *International Journal of Computer Applications* (0975–8887), 98(22), July 2014.

14. Syed Umar Amin, Kavita Agarwal, and Dr. Rizwan Beg, "Genetic Neural Network Based Data Mining in Prediction of Heart Disease Using Risk Factors" 978-1-4673-5758-6/13/$31.00 © 2013 IEEE.

15. D. Jain, "A Comparison of Data Mining Tools Using the Implementation of C4.5 Algorithm" *International Journal of Science and Research*, ISSN (Online), 3(8): 2319–7064, 2012.

Challenges in Internet of Things

4

Energy-Efficient Wearables

4.1 WEARABLE INTERNET OF THINGS

4.1.1 Wireless Body Area Network

With the advancements in wireless devices and communication systems in the medical field, Wireless Body Area Networks (WBANs) have promoted researchers to work on new things with an understanding thought. WBANs are progressively enhancing plausible results in various fields like health care, security, sports, entertainment, human motion, etc.

A WBAN is a compiled group of wireless sensors that can monitor an individual's physiological activities and actions, continuously, like the motion pattern of a human or their health status. This WBAN is also cited as Body Area Network (BAN) or Body Sensor Network (BSN) or even Medical Body Area Network.

The WBAN sensors are used in three ways: first, in which sensors are embedded inside a human's body; second, in which sensors are mounted over the surface but in a fixed position; and, third, in which sensors are accompanied within a device that can be easily carried in different positions.

DOI: 10.1201/9781003310341-4

4.1.2 WBAN Architecture

In common, WBAN architecture comprises the following three-tier communication:

A. **Intra-BAN communications**, in which communication takes place between the master node and wireless body sensors.
B. **Inter-BAN communications**, in which the master node communicates with personal devices like notepads, robots for home assistance, etc.
C. **Beyond-BAN communications**, where personal devices are connected to cyberspace.

Communications between different tiers are supported by several technologies, like Bluetooth, IEEE 802.15.4, and IEEE 802.15.6 etc (Figure 4.1). These technologies work out for BAN applications and acknowledge their maximum demands.

4.1.3 Body Sensors

The most essential elements for a WBAN are sensors and actuators. These sensors link the real, natural world to computerized systems. As these sensors are placed over a human body, with direct contact, and sometimes are even inserted, their dimensions and physical harmony with human tissues are crucial. This also leads to the research and making of new stuff (Figure 4.2).

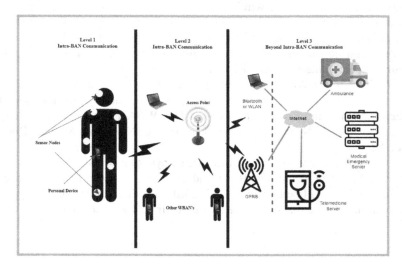

FIGURE 4.1 General Architecture for WBAN.

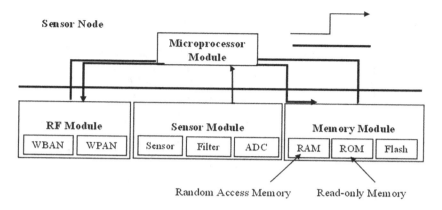

FIGURE 4.2   Components in a Sensor Node.

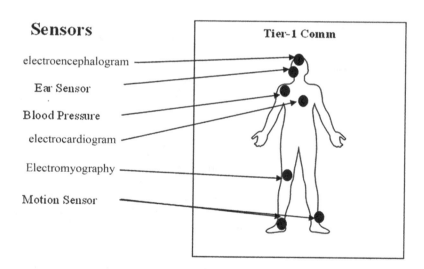

FIGURE 4.3   Body Sensors in WBAN.

An ordinary sensor node has a sensor, radio, and memory section. The sensor section comprises a sensor, an analogue-to-digital converter, and a filter. These sensors transform energy to analogue signals, which are band pass-filtered and loaded by Analog to digital converter (ADC) for more refining (Figure 4.3).

Some common body sensors are as follows (Table 4.1):

TABLE 4.1 Some Common Body Sensors and Purpose

S. NO.	SENSOR	PURPOSE(S)
1	Electroencephalogram (EEG) sensor	Little, tiny electrodes are adhered to at different positions of the scalp, in order to capture and assess the electrical activities happening inside the brain.
2	Electrocardiogram (ECG) sensor	This is used to investigate a heart's condition. This generates a pictorial transcript. Various heart medications, followed by practitioners, can also be observed and assessed for their performance.
3	Accelerometer/ Gyroscope	Different body poses and modes are identified and observed through this.
4	Electromyography (EMG) sensor	As muscles are controlled by nerves through electrical signals, any disorder causes an aberrant reaction. This sensor is used to measure those aberrant signals.
5	Glucometer	This one produces a numeric value that represents the quantity of glucose in the blood. Nowadays, glucose is observed through non-intrusive ways too.
6	Blood pressure	Blood pressure (systolic and diastolic, both) is estimated by using the vacillation method, through this non-intrusive sensor.
7	Pulse oximetry	A non-intrusive examination is done to check whether oxygen is normal or overloaded. A sensor is adhered to the patient's finger, lobe, or toe, by using a small clip. This sensor emits a light that penetrates through the skin, and as per the digested light of aerated haemoglobin and the whole amount of haemoglobin in the blood, the amount of oxygenated haemoglobin is declared.
8	Humidity and temperature sensors	This sensor gives an alarm, in a case, if a specified variation is observed in the body temperature and/or the humidity in an area, surrounding a person.
9	CO_2 gas sensor	This is used to measure and observe variations in the level of carbon dioxide and oxygen, during breathing.

4.1.4 WBAN Technologies

WBAN may use various and diverse technologies at different levels. These technologies include the following:

A. Bluetooth:

This standard communicates in a wireless field and that too within a limited range. This is proposed to implement an extreme level of security. Gadgets enabled with Bluetooth work in the 2.4 GHz ISM band, and the utmost rate of data transfer is 3 Mbps.

B. Bluetooth Low Energy (BLE):

This is an inherited preference after the Bluetooth standard, which uses very little power by performing small-duty cycle processes. This one was devised to hook up small devices with mobile terminals, wirelessly. They support a rate of data transfer up to 1 Mbps. In comparison to Bluetooth's seconds, synchronization in BLE takes place within a few milliseconds, so this is advised for applications demanding crucial latency, like alarm generators and emergency responders. This also enhances power saving.

C. ZigBee and 802.15.4:

ZigBee technology is broadly used from the low-energy surroundings for a wireless network. This is aimed for applications, operating on radio frequency which needs a lower rate of data transfer, extended battery time, and safe networking. The frequency bands in which these devices work are 868 MHz, 915 MHz, and 2.4 GHz. As in 2.4 GHz, several wireless systems work, and ZigBee faces interference. Also, the data rate is too low, only 250 Kbps, which makes it unsuitable for major and instantaneous applications. But yes, this is optimal for personal use, that is, a single patient.

D. IEEE 802.11:

This is a group of standards for wireless LAN. In this, users are either linked to an access point or kept in ad hoc style for surfing the Internet. This one is optimally suitable for transferring huge data by means of speedy wireless connectivity and also allows voice and video calls and streaming of video data, but the energy consumption is very high.

E. IEEE 802.15.6 9:

This is one of the earliest WBAN benchmarks, which handles numerous health and non-health applications. This also provides communications around and inside the body. Data transmission is managed in three diverse frequency bands:

a. The range of 400, 800, 900 MHz, and the 2.3 and 2.4 GHz bands form a **Narrowband**.

b. The highest 3.111.2 GHz is for **Ultra-Wideband** (UWB).

c. The range of 1050 MHz defines **Human Body Communication**. This one is headway in wearable wireless sensor networks as they support a broad range of data transfer rates, consume less energy, adequate 256 numbers of nodes for each BAN, and diverse node preferences in accordance with the application requirements. The maximum data rates can reach up to 10 Mbps although being tremendously low power.

This can deal with some body movements, like walking in a straight line between two points, but yes not appropriate for sitting, laying, footing up, jogging, swimming, and running. Also, it is unable to cope with the pressure of the newly coming applications, demanding a steep kind of audio or video.

F. UWB:

This technology supplies a high bandwidth and is employed for systems that communicate in a short range. This is the only trustworthy method of localization in an indoor environment, but because of its difficulty, it is inappropriate for wearable applications.

G. Wireless communications protocol stack (ANT) Protocol:

This is an up-and-coming benchmark for wellness and healthiness monitoring applications. This is a low-pace and low-powered code of behaviour and is carried up by a number of sensor makers.

H. Others:

The **Zarlink** is an extremely low-power technology that demands low frequency and a moderate rate of data transfer and is proper for medical embed applications. **RuBee active** sends data through lengthy wave magnetic signals and collects small-sized (128 bytes) data packets in a wireless LAN. Here a line of sight is not required for communication while performing operations. This provides effective transmission distance, strong defence, extremely low utilization of energy, steady function, and extended battery life. These all features do well for mobility-based medical care and observing and tracking a patient's health.

Proper and suitable radio technology for WBANs can be settled based on:

a. the particular demand of an application;

b. architecture level at which it will be set up.

Focussing on the requirements of WBAN, they can be listed as follows:

a. **Reliability**: High reliability is needed for health information sent by sensors.

b. **Latency**: Real-time transmission cannot stand for extended response time while handling emergency data, so performance assurance is required.

c. **Security**: Security and privacy are a big concern as systems are handling personal and critical data.

d. **Power Consumption**: Battery replacement is quite easy in comparison to other sensor networks; thus, there is very little concern on power usage, but this is limited to some scenarios only.

Now if the level of architecture is concerned, power, latency, and throughput can be concerned for intra-BAN communications that take place between a master node and the body sensors. For inter-BAN communications that take place between a master node and a single or more access points, collision and obstructions can appear over a mutual or common channel effortlessly so WLAN, Bluetooth, Zigbee, and cellular are possible. Beyond-BAN exchanges are necessary to facilitate approved healthcare human resources to access the subject's health information, remotely using a cellular network or the Internet.

4.2 ISSUES AND CHALLENGES IN WBAN

Although WBANs form a subgroup of general Wireless sensor network (WSNs), and even they share several familiar challenges, but still, a large number of inequalities survive between these two concerning the following:

A. Security:

This is a crucial issue as several keys are hacked using Denial of service (DoS) attacks, privacy contravention, and physical invasion. Providing safeguards against these hacks is difficult because of the limits on power consumption, data processing and transmission capability, and storage of the sensor nodes. These attacks may turn in severe results, so it should be considered with top preference [1, 2].

B. Energy Efficiency:

The sensor nodes in WBANs need to be very small, fine enough for making them wearable and implantable. The size specification puts a strict bar on the battery size of the nodes. Because these nodes cannot be recharged and replaced without surgery, it is expected from them to work for a long time. Therefore, these sensor nodes need to be very careful while consuming their energy. They consume energy during data collection, data processing, and data transmission [1–4].

C Localization:

Localization is an important feature for detecting the position of target sensor nodes which are distributed randomly in a network and for context discovery. The position of the sensors on the body surface or implanted, inside the body, including their orientation related to each other and to the body, mainly affects the strength of the signal in a WBAN. If the sensor is inadequately planted or displaced due to motion, it will give an incorrect estimation of a patient's location, and this will allow an attacker to transmit fake details. This way data quality can be compromised, so a reliable localization is highly required [5–7].

D. Interference:

Packet loss or collision happens when too many people wearing WBANs gather into an area, within the range of other WBANs, because the devices of the same WBANs do not communicate with the devices of other WBANs. This way, sent data does not reach its destination, so it is required to allocate multiple channels for reducing chances of collision [2, 7].

E. Sensor and System Integration:

Sensor integration reduces the weight and size of sensors, which makes them easy to wear and use. This way, manufacturing and maintenance costs can also be balanced, but it is quite difficult. System assimilation permits sensing through context that betters ease of use; approaches to openly accessible resources, turning into reduced operation cost; and also improves privacy and security [4, 6].

F. Different Layers, Different Faces:

Besides the above challenges, three different layers, physical, MAC, and network, have their challenges. The physical layer faces difficulties with rising temperature, dynamic topology, and the need for variable bandwidth. The MAC layer is supposed to deal with dynamic channel assignment, controlling packets and protocol overhead, synchronization, throughput, consistency or reliability, over-emitting, packet scheduling, error control, overhearing, calibration, quality of service (QoS), and multi-radio and multi-channel design. The network layer finds it difficult to work on optimum routing, mobility, traffic control, multi-path routing, and real-time streaming [4, 8].

4.3 LOCALIZATION IN WBAN

Traditional wireless sensor networks have already been equipped with a large number of protocols and algorithms. But, because of the exclusive characteristics and demands of WBAN applications, the suggested range for WSNs is not so suitable for WBANs. These differences can be clarified based on the following:

A. **Distribution, Arrangement, and Denseness**:
There are various points on which the number and distribution of sensor nodes depend. Generally, WSNs are deployed in unmanned areas, from where it becomes quite difficult for the operators to access them, and that is the main reason behind placing more nodes to compensate and make up, in case of any failures. WBAN nodes are deployed strategically over the human body surface and inside the body or are covered under clothes. Also, they do not engage duplicate nodes, unnecessarily to manage different kinds of failures. Denseness in WBANs is very low.

B. **Data Transfer Rate**:
WSNs are engaged mostly for monitoring events, happening at irregular intervals. Comparatively, WBANs are engaged in observing, monitoring, and recording humans' physiological activities, actions, and respective behaviours. This happens in a bit more periodic way, and thus, data transfer rates are relatively more stable.

C. **Latency**:
This need is prescribed by the functions and the operations. This may also be traded for enhanced reliability and energy efficiency. However, preserving energy is surely profitable. Batteries in BAN nodes can be replaced in a much easier way than in WSNs, where nodes are not physically reachable after

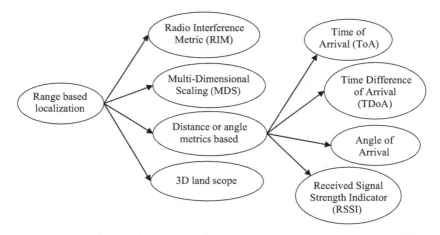

FIGURE 4.4   Range-based Location Tracking.

distribution. Thus, it becomes essential to magnify battery life in a WSN.

D. **Mobility**:

Generally, nodes in WSNs are treated as immobile, whereas BAN nodes or users may keep on moving around; that is why they experience the same mobility pattern. Localization has become a significant concern in WBANs because of the built-in features of sensor networks. Localization detects the location of targeted sensor nodes which are distributed randomly in an environment. These localization algorithms are broadly classified as:

A. **Range-based Localization**:

In this, information of location is received based on Time of Arrival, Angle of Arrival, Time – a Difference of Arrival, and Received Signal Strength Indication (RSSI). Once information is received, the location is estimated. These algorithms are good at accuracy but need additional range of information as they are either corrupted or faded by noise (Figure 4.4).

B. **Range-free Localization**:

In this, anonymous nodes collect connectivity-related information from anchors and then estimate location. Approximation Point in Triangle, Centroid, and Distance Vector Hop are used for collecting this information. They don't need any additional range of data, but they are not that much accurate (Figure 4.5).

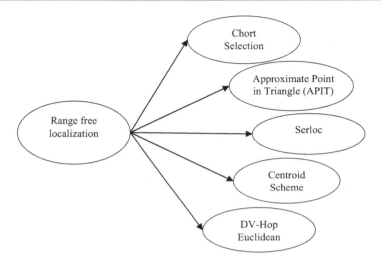

FIGURE 4.5   Range-free Location Tracking.

4.4 WBAN AND EARLIER STUDY

Former investigations referenced targeted on expensive apparatus, for channel-modelling functions [28] and for the affirmation of impulse radio based on-surface positioning [10, 29] for WBAN measures. One approach in [30] showed that unknown nodes estimate relative distance by counting number of hops and hop size (Table 4.2). In [31], non-linear systems used particle-filtering algorithm for localization purpose. This algorithm takes in information about radio signal and its strength from beacon messages from its neighbours to derive its location.

A method is proposed in [13] for estimating the position of sensors on the user's body, taking in account the motion data, using supervised and unsupervised learning approaches. It can earn 89% accuracy in finding a device's location. But the proposed method needs off-body processing and a minimum time of 30 minutes to capture motion data; therefore, this method is not suitable for real-time monitoring of the subject's body part. Further, the achieved accuracy highly depends on the measured limb.

One more study proposed automatic identification method, which enabled unassisted sensor nodes to monitor node locations, repeatedly [32]. But, this too failed in addressing the verification issue of the vertical location of the sensor.

TABLE 4.2 WBAN and Earlier Study

YEAR	APPROACH	RESULT(S)	NUMERICAL RESULT(S)	WORK NOT DONE/FUTURE WORK
2009 [9]	A spatial diversification, range-free localization algorithm was proposed, which combines several receivers in a body area sensor network (BASN) for location estimation.	The proposed RSSI method uses multiple receivers and improves location accurateness by beating the errors introduced by undersupplied antennas and contending fading in relation to space differences.	**Scenario A: (10)** Median Error <2m Deviation < 2.5m Largest Error <5.5m Average Error <=2.5m **Scenario B: (9)** Median Error ~=2m Deviation <6m Largest Error <7m Average Error <=3.3m	(a) Unable to contend with the error induced by shadowing, fading, and reflections. (b) Inaccurate measures of RSSI are not sieved while doing average computation. (c) Unable to accelerate the estimation process and resource optimization when using a large number of sensors.
2017 [10]	A mobile anchor-based localization algorithm is proposed which naturally determines the position of sensing nodes with the aid of the global positioning system (GPS) and uses trilateration rule for unknown nodes.	This model determines the position of sensing nodes with the aid of mobile anchors facilitated trilateration rule. Latency and throughput efficiency metrics in terms of percentage failure and localization error between the nodes are evaluated.	**Percentage Failure:** In comparison with 15m, for 45m, it is decreased by 93.88%, 48.28%, 11.76%, and 25% for anchor node's density 0.025, 0.1, 0.15, and 0.2, respectively. In comparison with 30m, for 45m, it is decreased by 49.28%, 8.69%, no change, and 13.33% for anchor node's density 0.025, 0.1, 0.15, and 0.2, respectively.	(a) The number of nodes and localization are not in a linear fashion, so it's quite difficult to identify the optimal number of nodes. (b) Although it is claimed that the proposed method also provides low cost and high energy efficiency, but it has not been shown or computed. Instead of this, it has been presumed that anchor nodes have enough energy in initial but energy consumption, and degradation in capacity has not been counted.

| 2012 [11] | This proposes evaluation and also compares the performance of localization methods for both inside and outside surroundings. For indoor environments, DV-hop algorithm, received signal strength (ROCRSSI), | (a) The particle filtering beats all others for an indoor environment, in terms of location accurateness for random and non-linear systems. (b) A mathematical equation to calculate Euclidean distance for the deterministic fingerprinting location tracking method is proposed. | **Localization Error:** In comparison with 10m, for 20m, it is decreased by 14.63%, 13.33%, 21.28%, and 13.33% for anchor node's density 0.025, 0.1, 0.15, and 0.2, respectively. In comparison with 15m, for 20m, it is decreased by 10%, 3.51%, 4.65%, and no change for anchor node's density 0.025, 0.1, 0.15, and 0.2, respectively **Estimation error:** DV-hop = 3.6039 Kalman filter = 0.92679. Particle filter l = 0.62056. *No numerical result is shown for outdoor environment-based location tracking.* | (a) Although the particle filter-based location tracking technique is best, it has a major drawback because of its higher computational overheads. (b) Construction of fingerprints is time-consuming (although no numerical result is shown for this), but no mechanism is suggested to reduce this time. (c) Energy efficiency is an important factor, but it is not shown in the whole study. |

(Continued)

TABLE 4.2 Continued WBAN and Earlier Study

YEAR	APPROACH	RESULT(S)	NUMERICAL RESULT(S)	WORK NOT DONE/FUTURE WORK
	particle filtering, and Kalman filtering–based tracking methods are considered. And for outdoor environments, GPS-based tracking and Global System for Mobile communication (GSM)-based tracking methods like Cell ID–based, deterministic fingerprints–based, probabilistic fingerprints–based, and Hidden Markov Model are proposed.	(c) For probabilistic fingerprinting, the area is divided into a number of cells, which helps with a reduced number of overheads as well as a reduced size of the fingerprint. (d) One more mathematical equation for steady-state probability for prediction is also proposed by using the Markov probability distribution function.		

(e) A hidden Markov GSM-based location tracking scheme performs best among all for an outdoor environment, in terms of location accurateness and computational overheads.

2018 [12]

A new method is proposed to compute and spot the arrangement of sensors devoid of using beacon nodes. The method also proposes location verification through differences in air pressure and RSSI measurements.

The proposed method recognizes a misplaced sensor, instantly. This is done through Barometric altimetry for vertical assessment and RSSI for horizontal verification.

(a) The proposed scheme is able to identify and position a human body within a range of 30 cm horizontally and 65 cm vertically and also identifies a limb.

(b) Two sample tests are performed, with three different transmitting powers.
Mean difference:
0dBm: 8.63
−5dBm: 4.14
−10dBm: 4.05
This shows that the higher the difference of mean, the higher the accuracy of position, and in this case, 0dBm beats the other two.

(a) This is applied only in an indoor environment that too without constraints and on limited sensors.

(b) No contextual information has been captured for monitoring body states.

(c) No work is suggested for detecting true node placement.

(d) Precision on location index is not counted.

(Continued)

TABLE 4.2 Continued WBAN and Earlier Study

YEAR	APPROACH	RESULT(S)	NUMERICAL RESULT(S)	WORK NOT DONE/FUTURE WORK
2011 [13]	An on-body device localization technique is proposed which uses acceleration data to capture motion data and runs two stages: unsupervised motion detection and supervised position assessment.	A method is discovered to estimate the positions of sensors placed on a person's body. This is done by capturing acceleration data from daily activities, such as walking. At first, the technique is detecting the time intervals when the user is in action, that is, walking through unsupervised learning, and then a support vector machine is used for pattern analysis and discovering the location of sensor devices on the body.	(a) Walking interval discovery: Highest 95% in foot area and lowest 89% in hand area is measured. (b) Location detection: Minimum, mean, and maximum overall classification precisions are 88%, 94%, and 100%, respectively. (c) The average classification accuracy is 89%.	(a) None activity except walking is considered. (b) Sensitivity of location and device's adaptability require improvement. (c) No significant classification is done for differing left and right limb. (d) Off-body processing is not concerned. (e) Minimum 30 minutes are required to capture motion information, that is, too much time and power consuming.

(Continued)

| 2014 [14] | This paper has evaluated a single hop and a two hop topology in terms of reliability, for three metrics packet deliverance, energy expenditure, and network's lifetime. | The results proved that off-body relays (2 hops) can be used to boost trustworthiness of data, lessen energy needs, and magnify network life, maximum. | *Packet Delivery Ratio:*
 1-hop:
 −20dBm: 40%, −10dBm: 86%, 0dBm: 97%
 2-hop off-body relay:
 −20dBm: 53%, −10dBm: 95%, 0dBm: 99%
 2-hop on-body relay:
 −20dBm: 43%, −10dBm: 91%, 0dBm: 99%
 Energy Cost Analysis:
 1-hop:
 −20dBm: 70mA, −10dBm: 14mA, 0dBm: 18mA
 2-hop off-body relay:
 −20dBm: 48mA, 10dBm: 24mA, 0dBm: 34mA
 2-hop on-body relay:
 −20dBm: 53mA, −10dBm: 25mA, 0dBm: 34mA

 Network lifetime:
 1-hop:
 −20dBm: 400H, −10dBm: 5000H, 0dBm: 9500H
 2-hop off-body relay:
 −20dBm: 2500H, −10dBm: 15000H, 0dBm: 11000H
 2-hop on-body relay:
 −20dBm: 1200H, −10dBm: 6500H, 0dBm: 5100H | It is an analysis of WBAN topology. Off-body relay is giving best lifetime because it is considered with unlimited energy, that is, no energy limits. What if a limit is put on energy? This point is missed. |

TABLE 4.2 Continued WBAN and Earlier Study

YEAR	APPROACH	RESULT(S)	NUMERICAL RESULT(S)	WORK NOT DONE/FUTURE WORK
2016 [3]	An RSSI-based device-free localization system is proposed to spot a body's presence in a defined area. This is observed by configuring received signal strength indicator (IRIS) motes and changing measures of signal strength.	The outcomes confirmed that spotting a body and tracing it are feasible within the range of 1.0m distance from LOS Path. The signal strength varies by 3.97dBm with a immobile body and from 10dBm to 15dBm with a moving body.	Scenario 1: Empty room Avg. RSSI: −48.95dBm Scenario 2: Person present on LOS for 120 sec Avg. RSSI: −52.92dBm Scenario 3: Person standing at different positions Avg. RSSI: −50dBm (at a distance of 0.2m to 0.8m) Avg. RSSI: −49dBm (at a distance of 1.0m to 2.0m) Scenario 4: Person moving across LOS Avg. RSSI: −67dBm (crossing LOS Y → Z) Avg. RSSI: −62dBm (returning Z → Y) A serious attenuation of 15.04dBm and 10.04dBm is observed here in the first and second cross, respectively.	(a) Only a single person is tracked. (b) The range of tracking is limited to 1.0m only. (c) Only tracking is done counting and recognition of multiple objects are not covered. (d) Other physical activities are not counted. (e) IRIS mote has very low battery power.

| 2017 [15] | This paper examines the functioning of intra-WBSN by putting Body Node Coordinators (BNCs) effectively. Total four sensor nodes are deployed on arm, head, foot, and the remaining last as a centroid for the other deployed three sensor nodes. Data is captured by bio-sensors and shipped further to BNC through a forwarder. This way, complexity level of processing data is reduced. | Performance is examined on factors like network stability, network life, lasting energy, throughput, and path loss. The investigation result shows that network stability is better at BNC1; network lifetime is better at BNC2, whereas throughput is the same at BNC1 and BNC2 and is better than other positions. BNC2 gives better results in the case of residual energy and path loss. So, BNC2 is the most appropriate position. | Network Stability: 6238 (BNC1) Network lifetime: 9973 (BNC2) Residual energy: .202 (BNC1 and BNC2) Throughput: $4.4*10^4$ (BNC2) Path-loss: 7200 rounds 380dB (BNC2) | (a) The human body is assumed to be in the standing position; no motion or activity is counted. (b) Even if motions are not counted, an energy-efficient scheme is required for postural movement. |

(Continued)

TABLE 4.2 Continued WBAN and Earlier Study

YEAR	APPROACH	RESULT(S)	NUMERICAL RESULT(S)	WORK NOT DONE/FUTURE WORK
2019 [16]	A self-adaptive guard band (SAGB) protocol is proposed, which adapts the guard time by using a time recompense algorithm as per clock's drift rate and makes sure that nodes will have less duty cycle and improved snooze time, with the aim of reducing energy consumption.	The results assure that the node sustains both snoozing status and harmonization with the coordinator for the duration of beacon transmission at the same time, thus reducing energy consumption.	<u>Lifetime (3 nodes):</u> GB MAC: 1170 SAGB MAC: 1400 Improvement: 16.4%. In addition, SAGB MAC reduces duty cycle and synchronization frequency and increases sleep time. <u>Duty cycle:</u> (Data generation time is 20s) GB MAC: 0.18% SAGB MAC: 0.07% Improvement: 51.4% <u>Energy consumption:</u> When a node transmits 4 data packets/second to the coordinator: GB MAC: 0.35 J SAGB MAC: 0.17 J Improvement: 51.4%	(a) Applicable only when beacon-based approach is followed. (b) As a new protocol is proposed, basic performance parameters like reliable data transfer, optimal route selection, and security are not covered.

| 2015 [17] | With the aim of extending the network's life, three different algorithms are proposed here for the effective placement of BNCs. These are as follows: Distance-aware iterative BNC placement algorithm (DBP-I) Distance-aware fixed BNC placement algorithm (DBP-F) Position-aware BNC placement (PBP) algorithm In addition to these, a new metric "E_{av}/d^n" is also proposed to quantify the expected lifetime of a node. | The proposed algorithms are diverse in nature, like requirements, and their formations also show the diverse level of energy efficiency and operational performance. PBP algorithm outperforms all. | No. of physical alterations (I)

Computational Complexity:
DBP-I: O(I)
DBP-F: O(I)
PBP: O(I)

Message Exchange Complexity:
DBP-I: O(I+1)
DBP-F: O(I)
PBP: O(I)

Energy Efficiency:
DBP-I: 36.8%
DBP-F: 41.8%
PBP: 47.45% | (a) Reliability and control message overheads are not concerned.
(b) The body is assumed in a standing position only, and no motion is considered. |

(Continued)

TABLE 4.2 Continued WBAN and Earlier Study

YEAR	APPROACH	RESULT(S)	NUMERICAL RESULT(S)	WORK NOT DONE/FUTURE WORK
2012 [18]	This proposed a locomotion technique based on multi-dimensional scaling to estimate various body gestures and their motions, by extracting information from 3D coordinate files. Unfortunately, the aimed method didn't satisfy the requirement.	The primary aim to model human body gestures and motions didn't succeed, so it was further improved by SVD reconstruction algorithm and putting on constraints, on calculated distances.	Average Position Error: LMDS: 217mm LMDS, with SVD and distance constraints: 167mm Improvement: 22% CDF: LMDS: 250mm LMDS, with SVD and distance constraints: 50mm	The communication between access points and the coordinator element is not discussed.

	Then again an improved method is proposed by combining an Single value decomposition (SVD) reconstruction algorithm and putting on distance constraints.			
2013 [19]	In this, two indoor localization algorithms, vector algorithm and matrix algorithms, are compared. These both use RSSI for estimation. A comparison is performed on the basis of orientation spots, access points, and accuracy of the localization process.	The localization implementation was analysed for various and diverse numbers of access points (AP), and reference points (RP), over 200 MHz and 400 MHz frequencies. This is concluded that the vector algorithm functions better for accuracy, cost, and effort.	Matrix algorithm: As more than 1 RP may have the same matrix, it faces ambiguity. Ambiguity grows more badly with an increasing number of RPs. Even by reducing APs, ambiguity increases. Vector algorithm: Localization improves by increasing RPs. Increased APs result in better performance. The growing number of reference points compensates for the decreased number of access points; thus, it is cost-effective.	The ambiguity problem need to be addressed.

(Continued)

TABLE 4.2 Continued WBAN and Earlier Study

YEAR	APPROACH	RESULT(S)	NUMERICAL RESULT(S)	WORK NOT DONE/FUTURE WORK
2018 [20]	An energy-efficient and competent routing concept is proposed here for settling down the consumed energy and QoS demands. The sink node takes care of routing paths and makes sure that sensors don't involve in any other sort of computations, except sensing. Here, the main focus is on delivering medical data in an effective manner, for both energy and QoS.	The projected concept is a destination-initiated routing, where the sink node checks for an optimal path according to the type of data packets. This is required to make sure effective use of sensors, resources, and also data transmission under specified conditions, like medical areas. The performance is assessed through simulations, and results have been compared with the conventional and Real-time traffic-differentiated Quality of service (TLQoS) approaches.	The proposed one has given impressive outcomes in terms of communication capacity, reaction time, and energy consumption. In terms of channel reliability rate: (a) Communication load: Lower, because nodes are freed from the selection phase, so there is no exchange of extra packets. (b) Reaction time: Lower, because the sensor nodes are picked up by the sink node. (c) Energy consumption: Lower, because sensor nodes are freed early and all later operations are carried out by the sink device.	(a) Sensor nodes are placed at fixed locations in a deterministic fashion. (b) Simple operation, sensing, is done. Additional and complex operations are not included, so it is quite limited.

| 2019 [21] | Every time there is a dilemma regarding the number of nodes that should be placed on the body of a human being. | Three schemes, SIMPLE, LAEEBA, and EENMBAN, are analysed here for a network with the number of nodes, varying from 6 to 10. The results are not uniform for any single deployment; it is varying for all, so it is quite difficult to decide how many numbers of nodes should be deployed. | In terms of communication frequency:
(d) Communication load: Lower always, because the transmission data is just only the sensed data amount.
(e) Reaction time: Lower, because the traded packet number is low as no additional data packets are exchanged.
(f) Energy consumption: Lower, because sensor nodes are set up already with routing paths, and this way energy is preserved.
(a) For stability period and network life, LAEEBA is performing better than SIMPLE, for 7- and 8-node groups, and SIMPLE performed fair for 6-, 9-, and 10-node groups.
(b) EENMBAN performed best for every group up to 1,500 rounds, without any node loss. | A uniform routing is needed for deciding upon a sufficient number of nodes to be placed on the body of a human being. |

(Continued)

TABLE 4.2 Continued WBAN and Earlier Study

YEAR	APPROACH	RESULT(S)	NUMERICAL RESULT(S)	WORK NOT DONE/FUTURE WORK
	This paper focuses on how the performance gets affected if the number of deployed nodes is varied.		(c) In the case of throughput, LAEEBA performed best for the 6- and 7-node groups, and EENMBAN showed better for 8-, 9-, and 10-node groups.	
2017 [22]	Several already available routing protocols are reviewed, and ATTEMPT, SIMPLE, and EERDT protocols are compared.	Three protocols for WBAN SIMPLE, ATTEMPT, and EERDT are studied and compared on the basis of their objectives, used routing technique, throughput, and issues related to them.	(a) EERDT continued for a higher number of cycles than both ATTEMPT and SIMPLE, as the first node expired after 5,000 rounds. (b) In the case of data delivered, EERDT delivers the highest number of packets from the cluster head to the base station. (c) With the increasing number of rounds, energy consumption by the nodes also increases. ATTEMPT loses its 50% energy after 2,000 rounds, whereas SIMPLE and EERDT manage up to 3,000 rounds.	QoS needs to be improved by securing the WBAN.

			(d) The network lifetime of EERDT is higher (after 5,000 rounds) instead of SIMPLE and ATTEMPT (after 1,000 rounds).	
			(e) As per the above results, EERDT is found to be more stable and reliable.	
2018 [23]	Problems like elderly fall detection, fall localization, and power consumed by sensor nodes are discussed here.	With the use of S-BFDA, an elderly fall is found out with a high level of accurateness in both line-of-sight (LOS) and non-line-of-sight (NLOS) surroundings. The ANN technique is proposed to localize the fall which achieves minimum localization error.	(a) As a result, the proposed S-BFDA achieves 100% and 92.5% fall detection accuracy for LOS and NLOS surroundings, respectively. (b) An indoor localization inaccuracy is made better by applying the ANN optimization, with an average absolute error of 0.0094 and 0.0454m for both LOS and NLOS surroundings, respectively.	(a) The energy consumption of FDS needs to be minimized. (b) The battery life of the FDS also needs to be extended. (c) The localization error, especially for indoor NLOS environments, need to be minimized. (d) The displacement of the patient needs serious attention.

(Continued)

TABLE 4.2 Continued WBAN and Earlier Study

YEAR	APPROACH	RESULT(S)	NUMERICAL RESULT(S)	WORK NOT DONE/FUTURE WORK
	A new sensor-based fall detection algorithm (S-BFDA) is proposed for fall detection. The fall localization is determined through an artificial neural network (ANN).	An energy-efficient technique, DDA, is projected to reduce power use and prolong battery life.	(c) The mean absolute error (MAE) in LOS is approximately five times healthier in comparison to NLOS surroundings.	
	A data-driven algorithm is used to minimize power consumption, as fall detection system (FDS) has limited energy resources, and the battery can be exhausted in a short time.		(d) The FDS can work for 62 days without replenishing the battery unlike in other implementations. (e) Finally, the proposed S-BFDA, ANN technique, and DDA outperform previous approaches for accurateness, MAE, and battery life.	

| 2016 [24] | Radio frequency is used for communication in WBAN because of its cheap framework setup and easy accessibility. As these waves are induced into the tissue, a high temperature, during radiation for too much longer duration, may injure the neighbouring tissues and can also lead to the growth of bacteria. This paper explores the existing thermal aware routing protocols. | The demand for heat management and existing thermal-aware routing protocols, Thermal Aware Routing Algorithm (TARA), Least Temperature Routing (LTR) protocol, and Adaptive Least Temperature Routing (ALTR) protocol, Hotspot Prevention Routing (HPR) algorithm, Thermal Aware Shortest Hop Routing (TSHR) algorithm, Least Total Route Temperature (LTRT), and Lightweight Rendezvous Routing (LRR) are discussed here. Also, a comparison of distinct thermal-aware routing protocols is done. | (a) TARA follows a retraction approach. A hotspot is set above a threshold value of temperature. Packets are simply withdrawn from the hotspot, without considering any shortest route.
(b) LTR sets hop count as a threshold and forwards the packet to the node having the least temperature and discards if hop count reaches the threshold.
(c) ALTR is an improved version of LTR and forwards the packet to the node having the least temperature and uses SHR in case of reaching a threshold value. | (a) An energy-efficient algorithm is required which uses lesser control messages and monitors continuously.
(b) A fault-tolerant system needs to be included so that it handles faulty nodes automatically.
(c) It is highly required to increase network lifetime. |

(Continued)

TABLE 4.2 Continued WBAN and Earlier Study

YEAR	APPROACH	RESULT(S)	NUMERICAL RESULT(S)	WORK NOT DONE/FUTURE WORK
			(d) HPR uses SHR for forwarding packets with the following cases: (i) The packet is directly sent if the destination node is on the next hop. (ii) If the temperature is below the threshold on the next hop, the packet is sent. (iii) A packet is forwarded to the coolest unvisited neighbour if the next hop reaches the threshold. Due to too much propagation of temperature data, HPR faces overhead issues. (e) TSHR is specifically for the applications that have delivery on high priority. Packets are retransmitted even if dropped.	

(Continued)

(f) LTRT finds out the shortest path with the least temperature from the source to the destination by using the Dijkstra algorithm. It doesn't create hotspots and avoids repetitious multiple hops. But this also faces overhead issues because of propagated temperature information to each node with a regular interval.

(e) LR algorithm divides nodes into small clusters and observes temperature increasing event. Once it reaches above threshold, it stops that node and informs others to start.

TABLE 4.2 Continued WBAN and Earlier Study

YEAR	APPROACH	RESULT(S)	NUMERICAL RESULT(S)	WORK NOT DONE/FUTURE WORK
2018 [25]	A new genetic approach combined scheme is proposed to discover the most favourable cluster head in order to minimize energy use and extend the network life to its maximum. The projected one has been compared with LEACH and DEEC.	The projected genetic algorithm is motivated from Darwinian Principle of Natural Selection and produces better results in comparison to LEACH and DEEC routing protocols. A random distribution considered with an assumption that sensors fixed on body are able to measure various dissimilar physiological activities. The results are observed for network life, group head creation, energy dissipation, and throughput.	(a) Network lifetime: The proposed scheme beats LEACH and DEEC algorithms, and the working of the projected protocol improves by 84.61% and 53.84%, respectively. (b) Count of Cluster Heads: The number of cluster heads formed in each cycle directly defines the life of the network. The proposed system forms more cluster heads, that is, 25 than the other two LEACH and DEEC, 11 and 9, respectively. (c) Energy Dissipation: The system is considered to be more efficient, if there is a decrease in value of energy dissipation. There is a decrease of −38.58 % and −63.61 % in dissipation value of the proposed one, when compared to LEACH and DEEC, respectively.	(a) Different body postures need to be considered. (b) Network availability is required for all times. (c) Lack of tracking methods may lead to the exchange of improper details to secure localization and data confidentiality, as both are required.

| 2016 [26] | This paper surveys concept and characteristics of WBAN, its communication standards, and deployment methods. This survey also discusses trustworthiness and fault forbearance issues for WBANs, along with coexistence, managing intrusion, and power use. | This paper outlined and discussed the up-to-date models for fault forbearance, reliability issues, and challenges like coexistence and interferences. The work proposed till date, for fault tolerance majorly, has centre of attention only on fault discovery, assuming that node replacement and isolation are the only recovery approaches, but in case of WBAN, it introduces a huge delay in response. | Data fusion is a first move in the direction of fault recovery. Data can be derived by a central node, available from other sensors, and performs necessary operations until isolation and replacement take place. | (d) Throughput: The proposed one gets an improved throughput by 18.94% and 40.50%, when compared to DEEC and LEACH, respectively. | (a) Privacy and security need to be included to increase the adoption level of WBANs. (b) New cross-layer interference mitigation schemes need to be introduced for different mobility and coexistence paradigms. (c) The highest degree of reliability with 0% error is required to make it feasible for use. (d) It is highly required to implement and install fault tolerance applications. |

(Continued)

TABLE 4.2 Continued WBAN and Earlier Study

YEAR	APPROACH	RESULT(S)	NUMERICAL RESULT(S)	WORK NOT DONE/FUTURE WORK
2016 [27]	A diffraction-based prototype is proposed, which illustrates the effect of numerous targets on the received signal strength indicator (RSS) field. Here, three new algorithms, Joint Maximum Likelihood (JML), Successive Cancellation (SC), and ML-RTI, are suggested for online positioning, using the median and the diversion of body-produced RSS perturbation.	A novel analytical prototype originated from diffraction hypothesis is developed to foresee the outcomes of two co-placed objects on both RSS mean and variance. The proposed prototype describes the fading inferred on each link of the network, by the targets. Also, new JML, SC, ML-RTI, and Radio tomographic imaging -least square estimation (RTI-LS) approaches for positioning multiple targets are tested in a real 2-target indoor environment and compared with the help of experimental data.	The accuracy is assessed by calculating mean of the execution results over 9 localization outcomes and two targets. Here, Diffraction Model is abbreviated as DM and Fingerprinting is abbreviated as FP. (a) JML: DM: 1.02m and FP: 0.61m (b) SC: DM: 1.18m and FP: 0.68m (c) RTI- ML: DM: 1.05m and FP: 0.51m (d) RTI-LS: DM: 1.11m and FP: 0.78m JML and Radio tomographic imaging- ML radio-tomographic imaging (RTI-ML) perform best. But RTI-LS performs worst because of not exploiting the variance information. In SC, the accurateness is restricted by solo target. An average accurateness is achieved in an area of 0.5–1 m.	(a) The double-target shadowing is approximated with a simplified additive approach. (b) Mismodelling caused by multipath is not considered in diffraction analysis. (c) It is required to combine all approaches with Bayesian tracing in order to control the ambiguities caused by multiple targets and multiple paths.

As mentioned in [33], RSSI became the most used technology in a distinct range of systems because of its cost efficiency, power efficiency, and easily accessible nature [33]. On the basis of empirical studies done in [34], an average positioning error of 50 cm is implemented by selecting the radio frequency and algorithm parameters cautiously. But still, the use of RSSI-based localization is not properly explored.

The study in [35] addressed the complications of the technologies used for localization and challenges of localization and also argued for the demand for accuracy by a number of diverse applications on different environments and platforms.

Detailed identification and observations are done on the currently existing low-power communication technologies [36]. This is done for supporting the accelerated and speedy evolution and distribution of WBAN structures. This chapter majorly focused on observing aged or chronically ill subjects in a residential environment, remotely.

To consolidate and bring together the WBAN technologies, a nomenclature is proposed in [37], departing from the adequate terminology. This helps by simplifying the methods of seeking and indicating the information related to these technologies, reducing time and advancing the process of understanding key concepts of WBAN. By inspecting the existing literature study, one can easily recognize that various authors use distinct terminologies for a sole concept and sometimes a single terminology for distinct concepts. This creates disorientation and bars proper comparisons between systems.

A device-free localization system's performance, in terms of accuracy, degrades with the increase in interferences for an indoor environment. Locating an indoor user and tracking it on the basis of radio signals are not so easy because several objects like floors, fences, surrounding divider boundaries, and human bodies inside an enclosed area lead to a complicated pattern of attenuation and fading of the signals [38].

An energy-accumulating-based BAN is proposed for active health, and an optimum resource allotment strategy is discussed to the better energy efficiency of BAN, based on Time division multiple access (TDMA) [39]. There is no single algorithm that is suitable for all types of situations. As per the utilization, algorithms are classified as derivative-predicated algorithms, derivative-free algorithms, and bio-mimetic algorithms. The first two classical algorithms use either Hessian matrix-predicated methods or gradient-predicated methods, whereas bio-mimetic algorithms use pattern matrix-predicated methods, which provide arbitrary results for different situations [40].

4.4.1 What Is Not Being Done?

1. Localization accuracy is an important concern as there is no method to sieve improper RSSI measurements, reducing errors caused by shadowing and fading.
2. The location precision factor has not yet been calculated.
3. Every time it is a concern about location estimation only, but what about verification of that estimated location? Sensor misplacement recognition has not been done fully, like whether a sensor is on the body or is on the table, so it is required to implement it at a large scale.
4. Multi-target tracking is still pending, limited to only two to three targets.
5. There is no sufficient number of nodes yet declared that can be placed over a human body.
6. Different body postures for a moving human body are not yet tracked.
7. Energy consumption, minimum or maximum time to capture any information, has not yet been handled.
8. Network lifetime matters no doubt, but what about fault tolerance? This is a serious concern, especially if WBAN is implemented for medical applications.
9. Real-time monitoring conditions and a wide range of contextual information are still a challenge.

4.5 APPLICATIONS

Like other sensor networks, WBANs also have applications in various areas such as medical, entertainment, lifestyle, and military [41]. WBAN has opened new doors in the medical sector with a large range of possibilities, out of which some are as follows:

1. **Monitoring Health Remotely**: WBAN offers monitoring status of a subject's vital organs, remotely. This information can be stored remotely or through the control unit, and then further analysis can be done. This improves the efficiency of doctor-patient activities.
2. **Telemedicine**: WBAN helps the patients to consult with doctors online, sitting at a remote location, and can get e-prescription as per the information of their medical condition.

3. **Living through Assistance**: This is the most fascinating thing offered by WBAN. In this, physiological data is measured from a patient's body through body sensors and then transmitted to a specific medical centre server or control unit on a regular interval basis. This way a patient can get continuous assistance and support from the hospital, without even staying at the hospital. This is highly useful for older age and handicapped people. They can send an alarm notification in case of any emergency [42].

4. **Rehabilitation and Therapy**: This is a progressive course that utilizes feasible amenities to cure any undesired movement or gesture behaviour in order to achieve an ideal posture. A patient who has already faced a collapse is observed carefully and amended so as to maintain a proper movement pattern.

4.6 LIMITATIONS AND FUTURE SCOPE

Limitations for this earlier work are, in general, as follows:

1. As the earlier work has been limited to identification and verification of misplaced sensors in an indoor environment only, usage and applicability will also be restricted in a specific range.
2. The earlier work will not be dealing with security issues, specifically.
3. The earlier work will be implemented for monitoring heartbeat rate only, and other sensing parameters will not be considered.

From the perspective of future work, the proposed work can be implemented for an outdoor environment or a crowded area. In addition, security measures can be implemented. Another type of contextual information can be added to cover a better range of data.

REFERENCES

1. S. Furqan Qadri, S. Afsar Awan, Mh. Amjad, M. Anwar, and S. Shehzad, "Applications, Challenges, Security of Wireless Body Area Networks (WBANs) and Functionality of IEEE 802.15.4/Zigbee" *Science International* (Lahore), 25(4): 697–702, ISSN 1013–5316; CODEN: SINTE 8 697, 2013.

2. Y. Zhou, "Energy Efficient Wireless Body Area Network Design in Health Monitoring Scenarios" Electrical and Computer Engineering, The University of British Columbia (Vancouver), March 2017 ©Yang Zhou, 2017.

3. S. Shukri, L. M. Kamarudin, Goh Chew Cheik, R. Gunasagaran, A. Zakaria, K. Kamarudin, S. M. M. S Zakaria, A. Harun, and S. N. Azemi, "Analysis of RSSI-based DFL for Human Detection in Indoor Environment using IRIS mote" 2016 3rd International Conference on Electronic Design (ICED), August 11–12, 2016, Phuket, Thailand, 978-1-5090-2160-4/16 ©2016 IEEE.

4. G. K. Ragesh, and Dr. K. Baskaran, "An Overview of Applications, Standards and Challenges in Futuristic Wireless Body Area Networks" IJCSI International Journal of Computer Science Issues, 9(1), No 2, ISSN (Online): 1694–0814 www.IJCSI.org, January 2012.

5. S. Al-Janabi, I. Al-Shourbaji, Mh. Shojafar, and S. Shamshirband, "Survey of Main Challenges (Security and Privacy) in Wireless Body Area Networks for Healthcare Applications" Egyptian Informatics Journal, 18: 113–122 (http://dx.doi.org/10.1016/j.eij.2016.11.001) 1110–8665/_ 2016 Production and hosting by Elsevier B.V. on behalf of Faculty of Computers and Information, Cairo University, 2017.

6. E. Jovanov, and A. Milenkovic, "Body Area Networks for Ubiquitous Healthcare Applications: Opportunities and Challenges" Springer Science + Business Media, LLC 2011 Journal of Medical Systems, 35: 1245–1254, DOI 10.1007/s10916-011-9661-x Published online: 17 February 2011.

7. B. Johny, and A. Anpalagan, "Body Area Sensor Networks: Requirements, Operations, and Challenges" March/April 2014 0278–6648/14/$31.00©2014 IEEE DOI 10.1109/MPOT.2013.2286692.

8. A. Sangwan, and P. Pratim Bhattcharya, "A Study on Various Issues in Different Layers of WBAN" International Journal of Computer Applications (0975–8887), 129(11): 24–28, November 2015.

9. C. Guo, J. Wang, R. Venkatesha Prasad, and M. Jacobsson, "Improving Accuracy of Person Localization with Body Area Sensor Networks: Experimental Study" 978-1-4244-2309-5/09/$25.00 ©2009 IEEE.

10. M. G. Kavitha, and S. Sendhinathan, "Body Area Network with Mobile Anchor Based Localization" Cluster Compute© Springer Science + Business media 2017 DOI 10.1007/s10586-017-1175-y.

11. O. ur Rehman, N. Javaid, A. Bibi, and Z. Ali Khan, "Performance Study of Localization Techniques in Wireless Body Area Sensor Networks" IEEE 11th International Conference on Trust, Security and Privacy in Computing and Communications, 2012.

12. H. Wang, Y. Wen, and D. Zhao, "Location Verification Algorithm of Wearable Sensors for Wireless Body Area Networks" Technology and Health Care, 26: S3–S18 DOI 10.3233/THC-173812, IOS Press, 2018.

13. A. Vahdatpour, N. Amini, and M. Sarrafzadeh, "On-body Device Localization for Health and Medical Monitoring Applications" Proceedings of the IEEE International Conference on Pervasive Computing and Communications, IEEE 2011: 37–44.

14. F. Di Franco, I. Tinnirello, and Y. Ge, "1 Hop or 2 Hops: Topology Analysis in Body Area Network" 978-1-4799-5280-9/14 ©2014 IEEE.

15. T. Rashid, S. Kumar, and A. Kumar, "Effect of Body Node Coordinator (BNC) Positions on the Performance of Intra-Body Sensor Network (Intra-WBSN)" 2017 4th International Conference on Power Control and Embedded Systems (ICPCES), DOI: 10.1109/ICPCES.2017.8117613 978-1-5090-4426-9/17©2017 IEEE.

16. T. Bai, J. Lin, G. Li, H. Wang, P. Ran, Z. Li, Y. Pang, W. Wu, and G. Jeon, "An Optimized Protocol for QoS and Energy Efficiency on Wireless Body Area Networks" Special Issue on Software Defined Networking: Trends, Challenges and Prospective Smart Solutions Peer-to-Peer *Network Applications*, 12: 326–336, DOI 10.1007/s12083-017-0602–4 © Springer, 2019.

17. Md. T. Ishtaique ul Huque, K. S. Munasinghe, and A. Jamalipour, "Body Node Coordinator Placement Algorithms for Wireless Body Area Networks" *IEEE Internet of Things Journal*, 2(1), DOI 10.1109/JIOT.2014.2366110, February 2015.

18. M. Mhedhbi, Mh. Laaraiedh, and Bernard Uguen, "Constrained LMDS Technique for Human Motion and Gesture Estimation" 978-1-4673-1439-8/12 ©2012 IEEE.

19. H. Obeidat, R. A Abd-Alhameed, J. M Noras, S. Zhu, T. Ghazaany, N. T. Ali, and E. Elkhazmi, "Indoor Localization Using Received Signal Strength" 978-1-4799-3525-3/13 ©2013 IEEE.

20. N. Yessad, and M. Omar, "Reliable and Efficient Data Communication Protocol for WBAN Based Healthcare Systems" *International Journal of E-Health and Medical Communications*, 9(2), DOI: 10.4018/IJEHMC.2018040102 Copyright © 2018, IGI Global, April-June 2018.

21. S. Ahmed, N. Sadiq, K. Sadiq, N. Javaid, and M. Ali Taqi, "Node Density Analysis for WBAN Schemes in Terms of Stability and Throughput" Recent Trends and Advances in Wireless and IoT-enabled Networks, EAI/Springer Innovations in Communication and Computing, https://doi.org/10.1007/978-3-319-99966-1_23 © Springer Nature Switzerland AG 2019.

22. J. Anand, and D. Sethi, "Comparative Analysis of Energy Efficient Routing in WBAN" 3rd IEEE International Conference on "Computational Intelligence and Communication Technology" (IEEE-CICT 2017) 978-1-5090–6218-8/17/$31.00 ©2017 IEEE.

23. S. Kamel Gharghan, S. Latteef Mohammed, A. Al-Naji, M. Jawad Abu-AlShaeer, H. Mahmood Jawad, A. Mahmood Jawad, and J. Chahl, "Accurate Fall Detection and Localization for Elderly People Based on Neural Network and Energy-Efficient Wireless Sensor Network" *Energies*, 11, 2866, DOI:10.3390/en11112866 www.mdpi.com/journal/energies, 2018.

24. N. Muhammad Shehu, and M. Muhammad Adam, "A Survey on Thermal Aware Routing Protocols in WBAN" Proceeding of International Conference on Emerging Technologies in Engineering, Biomedical, Management and Science [ETEBMS-2016], 5–6 March 2016.

25. A. Umare, and Dr. P. Ghare, "Optimization of Routing Algorithm for WBAN Using Genetic Approach" 9th ICCCNT 2018 July 10–12, IISC, Bengaluru, India IEEE – 43488.

26. M. Salayma, A. Al-Dubai, I. Romdhani, and Y. Nasser, "Wireless Body Area Network (WBAN): A Survey on Reliability, Fault Tolerance, and Technologies Coexistence" http://dx.doi.org/10.1145/0000000.0000000 © 2016 ACM 1544–3558/2010/05-ART1.

27. M. Nicoli, V. Rampa, S. Savazzi, and S. Schiaroli, "Device-free Localization of Multiple Targets" 24th European Signal Processing Conference (EUSIPCO) 978-0-9928-6265-7/16/$31.00 ©2016 IEEE.

28. SL. Cotton, R. D'Errico, and C. Oestges, "A Review of Radio Channel Models for Body Centric Communications" *Radio Science*, 49(6): 371–388, 2014.

29. N. Yessad, S. Bouchelaghem, F. S. Ouada, and M. Omar, "Secure and Reliable Patient Body Motion Based Authentication Approach for Medical Body Area Networks" *Pervasive and Mobile Computing*, 42: 351–370 ©2017 Elsevier B. V., December 2017.

30. S. Lee, D. Lee, and C. Lee, "An Improved DV-Hop Localization Algorithm in Ad Hoc Networks" ICUIMC '10, Proceedings of the 4th International Conference on Ubiquitous Information Management and Communication, ACM New York, NY, USA © 2010 ISBN: 978-1-60558-893-3 doi>10.1145/2108616.2108698.

31. H. Ren, M. Q. –H. Meng, and C. Hu, "Tracking Service for Mobile Body Sensor Networks Based on Transmission Power Aware Medium Access Control" Proceedings of the 2007 IEEE International Conference on Mechatronics and Automation.

32. G. Lo, S. González-Valenzuela, and V. C. M. Leung, "Automatic Identification and Placement Verification of Wearable Wireless Sensor Nodes Using Atmospheric Air Pressure Distribution" Consumer Communications and Networking Conference (CCNC), IEEE 2012: 32–33.

33. Q. Luo, Y. Peng, J. Li, and X. Peng, "RSSI-based Localization through Uncertain Data Mapping for Wireless Sensor Networks" *IEEE Sensors Journal*, 16(9): 3155–3162, 2016.

34. A. Awad, T. Frunzke, and F. Dressler, "Adaptive Distance Estimation and Localization in WSN using RSSI Measures" Proceedings of the 10th Euromicro Conference on Digital System Design Architectures, Methods and Tools, IEEE 2007; 471–478.

35. [K. Pahlavan, P. Krishnamurthy, and Y. Geng, "Localization Challenges for the Emergence of the Smart World" Special Section On Challenges For Smart Worlds, Digital Object Identifier 10.1109/ACCESS.2015.2508648 Pg. 3058, volume 3 2169–3536 2015 IEEE, current version January 12, 2016.

36. Mh. Ghamari, B. Janko, R. Simon Sherratt, W. Harwin, R. Piechockic, and C. Soltanpur, "A Survey on Wireless Body Area Networks for eHealthcare Systems in Residential Environments" *Sensors*, 16: 831; Digital Object Identifier: 10.3390/s16060831 www.mdpi.com/journal/sensors, 2016.

37. A. Fajardo, J. Fernando, and R. de Sousa, "A Taxonomy for Learning, Teaching, and Assessing Wireless Body Area Networks" VII Latin American Symposium on Circuits and Systems (LASCAS) 2016 ISBN 978-1-4673–7835-2/16/$31.00©2016 IEEE.

38. N. Pirzada, Y. Nayan, F. Hassan, F. Subhan, and H. Sakidin, "WLAN Location Fingerprinting Technique for Device-free Indoor Localization System" 2016 3rd International. Conference On Computer and Information Sciences (ICCOINS) 978-1-5090-2549-7/16/©2016 IEEE.

39. Y. Hao, L. Peng, H. Lu, Mh. Mehedi Hassan, and A. Alamri, "Energy Harvesting Based Body Area Networks for Smart Health" *Sensors*, 17: 1602, DOI: 10.3390/s1707 1602, 2017.

40. R. Sharma, H. Singh Ryait, and A. Kumar Gupta, "Wireless Body Area Network – A Review" *Research Cell: An International Journal of Engineering Sciences,* 17 ISSN: 2229–6913 (Print), ISSN: 2320–0332 (Online) http://www.ijoes.vidya-publications.com ©2016 Vidya Publications, January 2016.

41. Md. Taslim Arefin, Mh. Hanif Ali, and A. K. M. Fazlul Haque, "Wireless Body Area Network: An Overview and Various Applications" *Journal of Computer and Communication*, 5: 53–64, Scientific Research Publishing ISSN 2327–5277, 2017.

42. S. Sonone, and V. Shrivastava, "Study of Process of Localization and Methods of Localization in Wireless Body Area Networks (WBANs)" *International Journal of Computer Science and Information Technologies*, 5(6): 7710–7714, 2014.

Cognitive Internet of Things

5

Rainfall Prediction for Effective Farming

5.1 FARMING AND COGNITIVE INTERNET OF THINGS

Prediction is a procedure of assessing or forecasting the foreseeable future that varies depending on past and nearby data. Predictions offer information about the imminent future actions and their implications for the management. This might not reduce the difficulties and the hesitancy of the future. Nevertheless, it increases the self-reliance of the management to craft imperative decisions. Forecasting is the foundation of premising. Forecasting uses various statistical data. Consequently, it is also called Statistical Analysis [1]. The significance of forecasting involves the following points:

- Forecasting provides reliable and relevant information about the present and past events and the probable future measures. This is very essential for sound planning.
- It gives self-belief to managers for making imperative decisions.
- It is the source for making planning grounds.
- It keeps managers alert and active to face the challenges of future measures and the changes in the atmosphere.

DOI: 10.1201/9781003310341-5

The limits of the prediction include the following points:

I. Analytical data gathering about the current, historical, and future involves plenty of time and capital. Subsequently, executives will be required to strike a balance between the expenditure of predicting and its refund. Nearly all of the small- and medium-sized businesses do not make the forecasting on account of the high price.

II. Predicting tasks will only be able to estimate future actions. It is not possible to promise that these measures shall take place in the future. Prolonged-term prediction is going to be less accurate by comparison with the short-range forecast.

III. Data forecast will be based on confident expectations. If these statements are inaccurate, the predictions will be inaccurate. Forecasting is reliant on previous measures. However, the past might not necessarily restate itself [2].

IV. Prediction needs appropriate skills and decisions on the role of managers. Predictions might get inaccurate due to awful judgement and abilities on the part of some of the managers. Therefore, predicting data is subject to human mistakes.

The difficulties originating from the natural resource constraints, growing disintegration of holdings, recurring climatic variations, growing input costs, and then post-harvest deficits create an immense challenge to enduring agricultural growth.

5.1.1 Gross Domestic Product and Agriculture

Since the instigation of financial reforms in 1991, progression in agricultural gross domestic product (GDP) has shown in height volatility. It has fluctuated from 4.8% per annum in the Eighth Five-Year Plan (1992–1996) to a low of 2.4% during the Tenth Plan (2002–2006) before rising to 4.1% in the Eleventh Plan (2007–2012), as shown in Figure 5.1. From Figure 5.1, we can conclude that most small and marginal prosperities make this high volatility even more troublesome for policymakers, as small and marginal farmers are exceedingly vulnerable to adversative climatic conditions.

From now on, rainfall prediction is the vital research area that might be helpful for India's GDP; for the policymaker it may be less troublesome in the diverse condition.

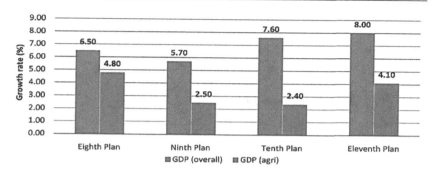

FIGURE 5.1 Agricultural Growth Rate during Different Plan Periods.
Source: Central Statistics Office (CSO)

5.2 MACHINE LEARNING MODEL FOR RAINFALL PREDICTION

In the prediction model (Figure 5.3), we have applied different machine learning classifiers [3–11], as we know that applying machine learning to any data needs a priori knowledge, like the type of data, size of data, and missing values, and the dataset can be categorized in two ways, as depicted in Figure 5.2.

5.2.1 Decision Tree Classifier

Decision Tree Classifier: The decision tree is a classic and natural model of learning. It is closely related to the fundamental computer science notion of "divide and conquer".

5.2.2 Support Vector Machine

Support Vector Machine (SVM): The objective of the support vector machine algorithm is to find a hyperplane in N-dimensional space (N – the number of features) that distinctly classifies the data points. To separate the two classes of data points, there are many possible hyperplanes that could be chosen. Our objective is to find a plane that has the maximum margin, that is, the maximum distance between data points of both classes. Maximizing the margin distance provides some reinforcement so that future data points

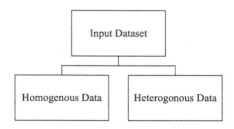

FIGURE 5.2 Types of Data.

can be classified with more confidence. Hyperplanes are decision boundaries that help classify the data points. Data points falling on either side of the hyperplane can be attributed to different classes. SVM maximizes the margin between the data points and the hyperplane. The loss function that helps maximize the margin is hinge loss as:

$$c(x, y, f(x)) = \begin{cases} 0, & \text{if} y * f(x) \geq 1 \\ 1 - y * f(x), & \text{else} \end{cases} \tag{1}$$

$$c(x, y, f(x)) = (1 - y * f(x)) \tag{2}$$

Equations 1 and 2 show the hinge loss function (Equation 1 can be represented as a function on Equation 2).

The cost is 0 if the predicted value and the actual value are of the same sign (Figure 5.3).

DecisionTreeTrain (data, remaining features)

1: guess ← most frequent answer in data // default answer for this data
2: if the labels in data are unambiguous then
3: return Leaf(guess) // base case: no need to split further
4: else if remaining features is empty then
5: return Leaf(guess) // base case: cannot split further
6: else // we need to query more features
7: for all f ε remaining features do
8: NO ← the subset of data on which f=no
9: YES ← the subset of data on which f=yes
10: score[f] ← # of majority vote answers in NO
11: + # of majority vote answers in YES
// the accuracy we would get if we only queried on f

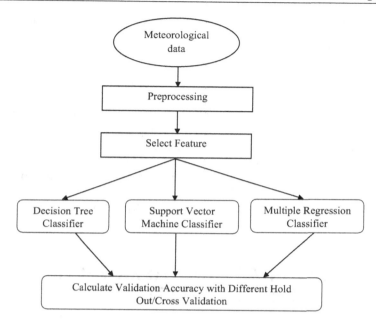

FIGURE 5.3   Prediction Model.

12: end for
13: f ← the feature with maximal score(f)
14: NO ← the subset of data on which f=no
15: YES ← the subset of data on which f=yes
16: left ← DecisionTreeTrain(NO, remaining features \ {f})
17: right ← DecisionTreeTrain(YES, remaining features \ {f})
18: return Node(f, left, right)
19: end if

5.2.3 Regression

Multiple Regression: Simple linear regression model has a continuous outcome and one predictor, whereas a multiple linear regression model has a continuous outcome and multiple predictors (continuous or categorical). A simple linear regression model would have the form:

$$y = \alpha + x\beta + \varepsilon$$

A multivariable or multiple linear regression model would take the form:

$$y = \alpha + x_1\beta_1 + x_2\beta_2 + \ldots + x_k\beta_k + \varepsilon$$

where y is a continuous dependent variable, x is a single predictor in the simple regression model, and $x_1, x_2, \ldots, x_k$ are the predictors in the multiple regression model.

In statistics, the mean squared error or mean squared deviation of an estimator (of a procedure for estimating an unobserved quantity) measures the average of the squares of the errors – that is, the average squared difference between the estimated values and what is actually estimated. Multiple linear regression can model more complex relationship, which comes from various features together. They should be used in cases where one variable is not evident enough to map the relationship between the independent and the dependent variable.

5.3 PRACTICAL APPROACH (MATLAB TOOL BOX)

For the implementation of our suggested algorithm, we have utilized Matlab 2019a. Rainfall dataset has been used from the Department of Agricultural Meteorology, Indira Gandhi Agricultural University, Raipur Station: Labhandi Monthly Meteorological Data: 2015. An example is as follows (Table 5.1):

Figure 5.4 shows the decision tree which is generated after applying the decision tree classifier to the input dataset (Table 5.2).

Matlab Code for Curve Fitting:

```
function [fitresult, gof] = createFit(x, y, z)
% Data for' fit:
% X Input : x
% Y Input : y
```

TABLE 5.1   Data Sample

YEAR	MAX_ TEMP	MIN_ TEMP	RELATIVE_ HUMIDITY1	RELATIVE_ HUMIDITY2	WIND_ VELOCITY	RAINFALL
2013	39.5	11.4	91	37	2.8	9.4
2013	31.9	14.4	85	33	2.9	2.2
2013	34.6	19.1	75	34	3.8	19.3
2013	39.3	23.1	73	35	7.2	51.4
2013	43.9	27.4	58	28	7.1	13.4

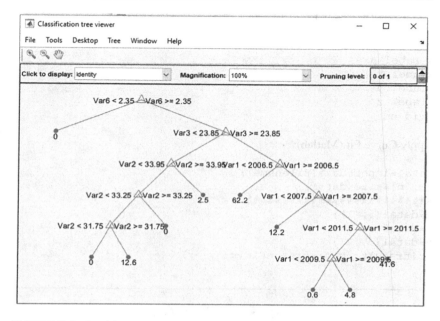

FIGURE 5.4   Decision Tree Generated Based on Input (Decision Tree Classifier).

TABLE 5.2   Accuracy

S. NO.	CLASSIFIER	ACCURACY (HOLD OUT)			
		0.2	0.3	0.4	0.5
1	Decision Tree Classifier	32	32	31	33
2	Naive Bayes Classifier	43	44	43	42
3	Multiple Regression	95	95.2	95	95.7

```
% Z Output: z
% Output:
% fit result : a fit object representing the fit.
% gof : structure with goodness-of fit info.
%% Fit: ' fit 1'.
[xData, yData, zData] = prepareSurfaceData(x, y, z);
% Set up fittype and options.
ft = fittype('poly11');
% Fit model to data.
[fitresult, gof] = fit([xData, yData], zData, ft);
% Plot fit with data.
figure('Name', fitness 1');
```

```
h = plot(fitresult, [xData, yData], zData);
legend(h, 'untitled fit 1', 'z vs. x, y', 'Location',
'NorthEast');
% Label axes
xlabel x
ylabel y
zlabel z
grid on
```

Apply Curve Fit (Matlab):

```
data1=importdata(filename);
[m, n]=size(data1);
%%%%%%%%%%%%%%%%%%%%%%%%%%%%%%
x=data1(:, 1);
z=data1(:, 3);
y=data1(:, 2);
[fitresult, gof]=createFit(x, y, z)
```

5.4 SUMMARY

India is an agricultural country, and the accomplishment or regret of the harvest and mineral water shortage in any given year is continually taken into consideration with the biggest concern. For the duration of the pre-summer, the region is thawed, provoking growing expansion and lower weight. Here's what inspires wind speed flow from the sea to the native land at low-down statures. Hereafter, we must have a prediction over rainfall; it will be beneficial for draught, farming system, etc.

In this chapter, we have used different machine learning algorithms to get a better precise rainfall prediction model. After experimental evaluation, we concluded that the multiple regression algorithm produces true positive rate as 95.8%.

REFERENCES

1. Hakan Tongal et al. *Phase-space Reconstruction and Self-exciting Threshold Modeling Approach to Forecast Lake Water Levels.* Springer-Verlag Berlin Heidelberg, 2013.
2. Andrew Kusiak et al. *Modeling and Prediction of Rainfall Using Radar Reflectivity Data: A Data-Mining Approach.* IEEE, 2013.

3. Pinky Saikia Dutta et al. / *Prediction of Rainfall Using Datamining Technique Over Assam.* IJCSE, 2014.
4. M. Kannan et al. *Rainfall Forecasting Using Data Mining Technique.* IJET, 2010.
5. Ravinesh C. Deo et al. *Application of the Extreme Learning Machine Algorithm for the Prediction of Monthly Effective Drought Index in Eastern Australia.* Elsevier, 2014.
6. Jae-Hyun Seo et al. *Feature Selection for Very Short-Term Heavy Rainfall Prediction Using Evolutionary Computation.* Hindawi, 2013.
7. Meghali A. Kalyankar, and Prof. S. J. Alaspurkar. Data Mining Technique to analyse Meterological Data IEEE Paper.
8. E. H. Habib, E. A. Meselhe, and A. V. Aduvala, "Effect of Local Errors of Tipping-bucket Rain Gauges on Rainfall–Runoff Simulations" *Journal of Hydrologic Engineering*, 13(6): 488–496, June 2008.
9. J. M. Sheridan, "Rainfall–streamflow Relations for Coastal Plain Watersheds" *Applied Engineering in Agriculture*, 13(3): 333–344, May 1997.
10. W. L. Crosson, C. E. Duchon, R. Raghavan, and S. J. Goodman, "Assessment of Rainfall Estimates Using a Standard ZR Relationship and the Probability Matching Method Applied to Composite Radar Data in Central Florida" *Journal of Applied Meteorology*, 35(8): 1203–1219, August 1996.
11. S. L. Neitsch, J. G. Arnold, J. R. Kiniry, and J. R. Williams, Soil and Water Assessment Tool User's Manual. Temple, TX: Soil Water Res. Lab., Agricultural Res. Service, Blackland Res. Center, Texas Agricultural Exp. Station, 2001, Ver. 2000.

Cognitive Internet of Things

6

Lake Level Prediction to Prevent Drought

6.1 DATA FORECASTING AND BOUNDARIES

Future forecasting is the process of predicting the future based on past and surrounding data. It provides forecast information about future actions to be taken and their consequences for the administration. This will not reduce the difficulties and contractions in the future. However, it increases the self-reliance of management to make the necessary decisions. Estimation is the basis of evidence. The reference uses various statistical data. Consequently, this is also known as statistical analysis [1,2]. The importance of assessment includes the following:

- Forecasting provides reliable and relevant information about the present and past events and probable future measures. This is very essential for sound planning.
- It gives self-belief to the managers for making imperative decisions.
- It is the source for making planning grounds.
- It keeps managers alert and active to face the challenges of future measures and the changes in the atmosphere.

DOI: 10.1201/9781003310341-6

Lake water elevation changes seasonally, for example, high in the extreme summer and low in the dried-out winter with sharp falling/rising limbs through typhoon events but not in a straightforward periodic mode, except for the seiche oscillation that occurs mostly in huge lakes. Effectual forecasting tools play a significant role in the studies of lake level measures. These can be used to replicate the lake water elevation diversity based upon the presence of measured data and forecast the probable responses under dissimilar scenarios, supporting management decisions of valuable water capital.

Forecasting of water elevations in the lakes [3] in cities that reach to city people for expenditure is mostly done physically and has the restraint that it cannot take into deliberation the past levels of water accessibility for forecasting the future levels' patterns that can help in better planning and can shun the shortage of water in cities to greater coverage. Forecasting approaches can be classified into various approaches [4,5]:

1. Qualitative Approach: In this approach, there is no use of any mathematical model due to the fact that the data available is not considered to be contributing to the future values (long-term forecasting).
2. Quantitative Approach: In this approach, historical data is available. It is based on analysis of historical data having the time series [H. Aksoy 2013] of a particular variable and other related time series. It also examines the cause-and-effect relationships of one type of variable versus other relevant variables.
3. Time Series Approach: In this approach, we have a single variable that keeps changing with time and whose future values are definitely related in some form to its past values.

6.1.1 Motivation

The availability and accessibility of near real-time water level data at multiple locations for each of the lakes were key components to the success of the operational forecasting system. The databases of historical water levels and wave heights were used to calibrate the hydrodynamic and wave models for each lake, while the near real-time water level data was used for data assimilation on the hydrodynamic models.

The ease of use and convenience of near real-time water level data at manifold locations for each one of the lakes were key constituents for the accomplishment of the prepared forecasting system. The datasets of chronological water levels and wave heights were worn to standardize the hydrodynamic and wave models for each lake, while the near real-time water level data was used for data absorption on the hydrodynamic models.

Accurate prediction of water level ambivalence is important in lake management owing to its remarkable impacts in a range of aspects. Here's the attempt to investigate and recommend a particular computer platform that may be used in surviving for the reservoir level. It's acknowledged that the system is without difficulty the scalability to incorporate the supplemental knowledge for the various reservoirs and continually perform professionally.

Forecasting of lake water level can be useful for:

- Evaluating the seriousness of the drought
- Forecasting where/when there is the danger of inundating
- Restricting the environmental impact of floods and droughts by allowing the government as well as other agencies to place the emergency reply from the plans into operation

It pinpoints vulnerable and risk-susceptible areas for floods and droughts in the entire province. Prediction is simply a prediction regarding the future values of data. Nevertheless, for the most part, predictive model forecasts are based on the assumption that the earlier is a proxy for the future. There are several different conventional models for predicting: exponential smoothing, regression, time series, and composite model forecasts, frequently comprising specialist forecasts. Regression analysis is a statistical technique to evaluate quantitative data to estimate model parameters and make forecasts.

6.2 ENSEMBLE PREDICTION MODEL

After reading several pieces of literature, we found some bottlenecks of the existing forecasting model, as lesser accuracy may lead to so many problems; hence, there is a need for high-accuracy machine learning prediction model. Secondly, as the amount of data is very high, optimizing machine learning classifier is required, and the noise present in the input dataset may affect the accuracy of the machine learning model. As dataset size is huge, different subsets of the dataset work better for different machine learning classifiers.

Henceforth, in this chapter, we have proposed the use of the weighted sum Ensemble machine learning approach, and we have proposed the use of preprocessing [6] for the removal of noise from the input dataset. Figure 6.1 will depict our proposed flow.

Ensemble method is a machine learning technique that combines several base models in order to produce one optimal predictive model. Ensemble methods are meta-algorithms that combine several machine learning techniques

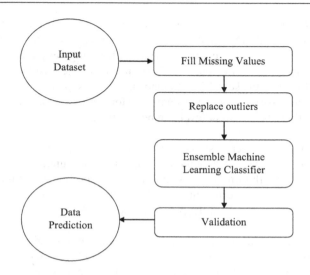

FIGURE6 1 System Flow.

into one predictive model in order to decrease variance (bagging) and bias (boosting) or improve predictions (stacking).

Proposed Algorithm

Step-1. //Read Input dataset from csv file.
    a. Check encoding (.csv,.txt etc)
    b. Read the sheet from csv
    c. Set top row as variable name

Step-2. //Convert data D to table array.

$$D_{rc} = \sum_{k=1}^{m} a_{rk} b_{kc}$$

Step-3. //Identification of missing values and replace with NaN.
    a. For each row in D

        if D_{ij}= Null
            D_{ij}=NaN
        end if
Step-4. Call **Fill_NaN (D_{ij})**.

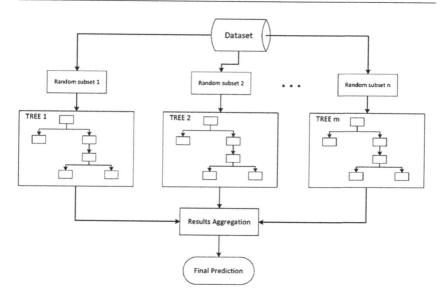

FIGURE 6.2   Ensemble Method.

Step-5. Call **Fill_Outlier (D_{ij}).**
Step-6. Call Ensemble
Step-7. Validate Prediction Model //Predicted data validation
Step-8. Calculate Accuracy, Sensitivity, Specificity, F-Score (Figure 6.2).

6.2.1 Preprocessing

Fill_NaN (D_{ij})

```
    Step-1. For each col (Ci) in D
            for each row in Ci
                    result= sum(isNaN(D.Ci))
                    if result==1
                            sort(Ci)
                            MM=Median(Ci)
                            D.Ci=MM
                    end if
            end for
    Step-2. end for
        return Dij
```

Fill Outlier (D_{ij})

Step-1. For each col (C_i) in D
    for each row in Ci
        Detect Outlier
        // Apply three sigma rule
        // Calculate Gaussian distribution using following equation

$$f(c, \mu, \sigma) = \frac{1}{\sigma\sqrt{2\pi}} e^{\frac{(c-\mu)^2}{2\sigma^2}}$$

// μ is Mean of data points
// σ is Standard deviation

$$\text{Calculate } Z = \frac{|c - \mu|}{\sigma}$$

        if Z==3 then
            return Z
        end if
    end for
Fills outlier with the lower threshold value for elements smaller than the lower threshold determined by 3-Sigma rule (Z). Fills outlier with the upper threshold value for elements larger than the upper threshold determined by 3-Sigma rule (Z).
Step-2. end for
return D_{ij}

6.2.2 Ensemble Model

Ensemble (Figure 6.2)

Step-1. Take any learner as a base learner that keeps all the distributions and allocates identical weight or attention to every observation.

Step-2. If prediction error occurs by 1st base learning algorithm, then put higher weight to sample having prediction error. Then, we put on the next base learning algorithm.

Step-3. Repeat Step 2 till the limit of the base learning algorithm is reached or higher accuracy is attained.

6.2.3 Ensemble Model Is Better Than Single Classifier

A learning algorithm can be viewed as searching a space H of hypotheses to identify the best hypothesis in the space. The statistical problem arises when the amount of training data available is too small compared to the size of the hypothesis space. Without sufficient data, the learning algorithm can find many different hypotheses in H that all give the same accuracy on the training data. By constructing an ensemble out of all of these accurate classifiers, the algorithm can average their votes and reduce the risk of choosing the wrong classifier. Figure 6.3 (top left) depicts this situation. The outer curve denotes the hypothesis space H. The inner curve denotes the set of hypotheses that all give good accuracy on the training data. The point labelled f is the true hypothesis, and we can see that by averaging the accurate hypotheses, we can find a good approximation to f.

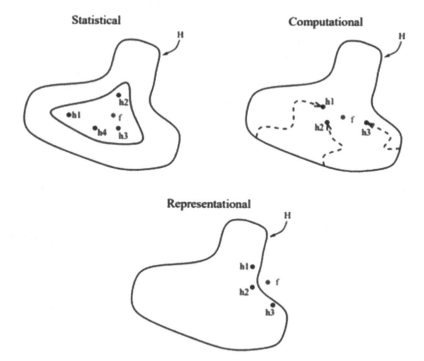

FIGURE 6.3 Three Fundamental Reasons Why an Ensemble May Work Better Than a Single Classifier [Thomas G Dietterich].

6.3 VALIDATION OF PREDICTION MODEL

Accurate predictions of water level fluctuation that results from hydro-meteorological variations and anthropogenic disturbances are needed for sustainable development and management of lake water usage.

Here, to visualize the expected result of our project and how we can predict the lake water level by applying different prediction models, we have demonstrated the process using Excel as follows:

Example input meteorological data, which is time series data, has been taken from: http://lakepowell.water-data.com/

6.4 SUMMARY

In this research, a portable and scalable prediction model is discussed for forecasting the water levels of lakes (Table 6.1). Accurate predictions of water level fluctuation that results from hydro-meteorological variations and anthropogenic disturbances are needed for sustainable development and management of lake water usage. Henceforth, greater accuracy is required for the prediction system; the proposed prediction model achieved an average accuracy of 98%.

TABLE 6.1   Accuracy of Lake Level with Different Holdouts

S. NO.	NAÏVE BAYES (NB)-1 (KERNEL)	NB-2 (MVN)	K-NEAREST NEIGHBOR (KNN)	SUPPORT VECTOR MACHINE (SVM)	DECISION TREE	PROPOSED
Holdout- 20						
1.	97.4	97.1	81	97.6	97.1	98.2
Holdout- 30						
2.	97.8	97.3	81.6	97.7	97.2	98.6
Holdout- 40						
3.	97.8	97.5	80.2	97.9	97.6	98.2
Holdout- 50						
4.	97.7	97.2	81.3	97.5	97.3	98.8
Average	**97.675**	**97.275**	**81.025**	**97.675**	**97.3**	**98.45**

REFERENCES

1. Mohammad Ali Ghorbani, Ravinesh C. Deo, Vahid Karimi, Zaher Mundher Yaseen, Ozlem Terzi, *Implementation of a Hybrid MLP-FFA Model for Water Level Prediction of Lake Egirdir*, Turkey, DOI 10.1007/s00477-017-1474-0 Springer, 2017.
2. Tengfei Hua, Jingqiao Maoa, Shunqi Panc, Lingquan Daid, Peipei Zhanga, Diandian Xua, Huichao Daia, *Water Level Management of Lakes Connected to Regulated Rivers: An Integrated Modeling and Analytical Methodology.* Elsevier, 2018.
3. Jalal Shiri, Shahaboddin Shamshirband, Ozgur Kisi, Sepideh Karimi, Seyyed M. Bateni, Seyed Hossein Hosseini Nezhad, and Arsalan Hashemi, *Prediction of Water-Level in the Urmia Lake Using the Extreme Learning Machine Approach.* Springer, 2016.
4. Jalal Shiri, Shahaboddin Shamshirband, Ozgur Kisi, Sepideh Karimi, Seyyed M. Bateni Seyed Hossein Hosseini Nezhad and Arsalan Hashemi, DOI 10.1007/s11269-016-1480 2016 Springer Science+Business Media Dordrecht, 2016.
5. Ozgur Kisi, Jalal Shiri, Sepideh Karimi, Shahaboddin Shamshirband, Shervin Motamedi, Dalibor Petkovic , and Roslan Hashim Elsevier *A Survey of Water Level Fluctuation Predicting in Urmia Lake Using Support Vector Machine with Firefly Algorithm*, 2015.
6. Prashant Shrivastava, S. Pandiaraj, and Dr. J. "Big Data Analytics in Forecasting Lakes Levels" Jagadeesan *International Journal of Application or Innovation in Engineering & Management (IJAIEM)*, 3(3), March 2014.

Appendix
MATLAB Implementation of Different Classifiers

The snippet of "Data_1.csv"

Location	Age	Annual Salary	Opinion
US	40	72000	not liked
Asia	25	48000	not liked
Africa	30	54000	not liked
Asia	35	61000	not liked
Africa	44		Liked
US	33	58000	Liked
Asia		52000	not liked
US	40	79000	Liked
Africa	55	83000	not liked
US	35	67000	Liked

1. Data Preprocessing

```
%%---------------Importing the dataset----------------------
%--------------------------Code--------------------------
data = readtable('Data_1.csv')
%%--------------Data Preprocessing--------------------------
%-------------Handling Missing Values--------------------
%-------------Method1: Deleting rows or column--------------
%--------------------------Code--------------------------
%
complete_data = rmmissing(data)
complete_data = rmmissing(data, 2)
% if second argumrnt given as 2then remove all col with
NAN value 1 then remove row with NAN
data = complete_data;
```

```
% -------------- Method 1.1: Deleting rows or columns
based on
%-------------------------Relative Percentage of missing---
%---------------------------Code--------------------------
restricted_missing = rmmissing(data, 'MinNumMissing', 3)
restricted_missing = rmmissing(data, 2,'MinNumMissing', 2);
restricted_missing = rmmissing(data, 2,'MinNumMissing', 3);
%
% data = restricted_missing;
%-------------Method2:Using Mean-------------------------
%-------------------------Code--------------------------
M_Age = mean(data.Age, 'omitnan');
U_Age = fillmissing(data.Age, 'constant', M_Age);
data.Age = U_Age;
%-------------Method3:Dealing with non-numeric data--------
%-------------------------Code--------------------------
data.Opinion = categorical(data.Opinion);
Freq_opinion = mode(data.Opinion);
Opinion = fillmissing(data.Opinion, 'constant',
cellstr(Freq_opinion));
data.Opinion = Opinion;
%%--------------Handling Outliers------------------------------
%--------------Method1:Deleting Rows----------------------
%-------------------------Code--------------------------
outlier = isoutlier(data.Age);
data = data(~outlier, :);
%--------------Method2:Filling Outliers--------------------
%-------------------------Code--------------------------
%
Age = filloutliers(data.Age, 'clip', 'mean')
data.Age = Age;
%% -------------- Categorical data-----------------
%%--------------Method1:Categorical data (noorder)----------
%-------------------------Code--------------------------
data = categorical_data_to_dummy_variables(data, data.
Location);
data.Location = [];
%--------------Method2:Categorical data (with order)----------
%-------------------------Code--------------------------
new_variable = categorical_data_to_numbers(data.
YearlyIncome, {'Average', 'High', 'Very High', 'Low'}, [2
3 5 1]);
data.YearlyIncome = new_variable
%%--------------Feature Scalling------------------------------
%--------------Method1:Standardization--------------------
%-------------------------Code--------------------------
```

```
stand_age = (data.Age - mean(data.Age))/std(data.Age)
data.Age = stand_age;
%--------------Method2:Normalization----------------------
%--------------------------Code--------------------------
normalize_age = (data.Age - min(data.Age)) / (max(data.
Age) - min(data.Age))
data.Age = normalize_age;
```

2. Decision Tree Classifier

```
clear all
%%--------------Importing the dataset----------------------
%--------------------------Code--------------------------
data = readtable('Data.csv');
%%%%--------------Classifying Data---------------------------
%%-------------Building Classifier--------------------------
%--------------------------Code--------------------------
classification_model = fitctree(data,
'Purchased~Age+Estimated Salary');
%%-------------Test and Trainsets--------------------------
%--------------------------Code--------------------------
cv = cvpartition(classification_model.NumObservations,
'HoldOut', 0.2);
cross_validated_model = crossval(classification_model,
'cvpartition', cv);
%%-------------Making Predictions for Testsets---------------
%--------------------------Code--------------------------
Predictions = predict(cross_validated_model.Trained{1},
data(test(cv), 1:end-1));
%%-------------Analyzing the predictions--------------------
%--------------------------Code--------------------------
Results = confusionmat(cross_validated_model.Y(test(cv)),
Predictions);
%%-------------Visualizing training setresults--------------
%--------------------------Code--------------------------
labels = unique(data.Purchased);
classifier_name = 'Decision Tree (Training)';
Age_range = min(data.Age(training(cv)))-1:0.01:max(data.
Age(training(cv)))+1;
Estimated_salary_range = min(data.EstimatedSalary(-
training(cv)))-1:0.01:max(data.
EstimatedSalary(training(cv)))+1;
[xx1, xx2] = meshgrid(Age_range, Estimated_salary_range);
XGrid = [xx1(:) xx2(:)];
predictions_meshgrid = predict(cross_validated_model.
Trained{1}, XGrid);
```

```
gscatter(xx1(:), xx2(:), predictions_meshgrid, 'rgb');
hold on
training_data = data(training(cv), :);
Y = ismember(training_data.Purchased, labels{1});
scatter(training_data.Age(Y), training_data.
EstimatedSalary(Y), 'o', 'MarkerEdgeColor', 'black',
'MarkerFaceColor', 'red');
scatter(training_data.Age(~Y), training_data.
EstimatedSalary(~Y), 'o', 'MarkerEdgeColor', 'black',
'MarkerFaceColor', 'green');
xlabel('Age');
ylabel('Estimated Salary');
title(classifier_name);
legend off, axis tight
legend(labels, 'Location', [0.45,0.01,0.45,0.05],
'Orientation', 'Horizontal');
%%--------------Visualizing testing set results----------------
%--------------------------Code--------------------------
labels = unique(data.Purchased);
classifier_name = 'Decision Tree (Testing)';
Age_range = min(data.Age(training(cv)))-1:0.01:max(data.
Age(training(cv)))+1;
Estimated_salary_range = min(data.EstimatedSalary
(training(cv)))-1:0.01:max(data.
EstimatedSalary(training(cv)))+1;
[xx1, xx2] = meshgrid(Age_range, Estimated_salary_range);
XGrid = [xx1(:) xx2(:)];
predictions_meshgrid = predict(cross_validated_model.
Trained{1}, XGrid);
figure
gscatter(xx1(:), xx2(:), predictions_meshgrid, 'rgb');
hold on
testing_data = data(test(cv), :);
Y = ismember(testing_data.Purchased, labels{1});
scatter(testing_data.Age(Y), testing_data.
EstimatedSalary(Y), 'o', 'MarkerEdgeColor', 'black',
'MarkerFaceColor', 'red');
scatter(testing_data.Age(~Y), testing_data.
EstimatedSalary(~Y), 'o', 'MarkerEdgeColor', 'black',
'MarkerFaceColor', 'green');
xlabel('Age');
ylabel('Estimated Salary');
title(classifier_name);
legend off, axis tight
legend(labels, 'Location', [0.45,0.01,0.45,0.05],
'Orientation', 'Horizontal');
```

3. KNN Classifier

```
clear all
%%--------------Importing the dataset----------------------
%---------------------------Code-------------------------
data = readtable('Data.csv');
%_____
%_____

%%%%---------------Classifying Data----------------------------
%%--------------Building Classifier-------------------------
%-------------------------------Code-------------------------
classification_model = fitcknn(data, 'Purchased~Age+Estim
atedSalary');
%please define your classifier here
%%--------------Test and Trainsets----------------------------
%---------------------------Code-------------------------
cv = cvpartition(classification_model.NumObservations,
'HoldOut', 0.2);
cross_validated_model = crossval(classification_model,
'cvpartition', cv);
%%-------------Making Predictions for Testsets--------------
%---------------------------Code-------------------------
Predictions = predict(cross_validated_model.Trained{1},
data(test(cv), 1:end-1));
%%-------------Analyzing the predictions--------------------
%---------------------------Code-------------------------
Results = confusionmat(cross_validated_model.Y(test(cv)),
Predictions);
```

4. Naive Bayes

```
clear all
%%--------------Importing the dataset----------------------
%---------------------------Code-------------------------
data = readtable(Data.csv');
%_____
%_____

%%%%---------------Classifying Data----------------------------
%% -------------- Building Classifier Different options
of Naïve Bayes---------------------------
%---------------------------Code-------------------------
classification_model = fitcnb(data, 'Purchased~Age+Estima
tedSalary');
classification_model_1 = fitcnb(data, 'Purchased~Age+Esti
matedSalary', 'Distribution', 'kernel');
```

```
%classification_model_1 = fitcnb(data, 'Purchased~Age+Est
imatedSalary', 'Distribution', 'mn');
%classification_model_1 = fitcnb(data, 'Purchased~Age+Est
imatedSalary', 'Distribution', 'normal');
%classification_model_1 = fitcnb(data, 'Purchased~Age+Est
imatedSalary', 'Distribution', 'mvmn');
%classification_model_1 = fitcnb(data, 'Purchased~Age+Est
imatedSalary', 'DistributionNames', {'kernel',
'kernel}');
% Please consult the video lectures for complete options
%%--------------Test and Trainsets--------------------------
%--------------------------Code--------------------------
cv = cvpartition(classification_model.NumObservations,
'HoldOut', 0.2);
cross_validated_model = crossval(classification_model,
'cvpartition', cv);
cross_validated_model_1 = crossval(classification_
model_1,'cvpartition', cv);
%%--------------Making Predictions for Testsets---------------
%--------------------------Code--------------------------
Predictions = predict(cross_validated_model.Trained{1},
data(test(cv), 1:end-1));
Predictions_1 = predict(cross_validated_model_1.
Trained{1}, data(test(cv), 1:end-1));
%%--------------Analyzing the predictions--------------------
%--------------------------Code--------------------------
Results = confusionmat(cross_validated_model.Y(test(cv)),
Predictions);
Results_1 = confusionmat(cross_validated_model_1.Y
(test(cv)), Predictions_1);
```

5. SVM Classifier

```
clear all
%%--------------Importing the dataset----------------------
%--------------------------Code--------------------------
data = readtable(Data.csv');
%%%%--------------Classifying Data----------------------------
%%-------------Building Classifier--------------------------
%--------------------------Code--------------------------
classification_model = fitcsvm(data, 'Purchased~Age+Estim
atedSalary');
classification_model_1 = fitcsvm(data, 'Purchased~Age+Est
imatedSalary', 'KernelFunction', 'polynomial');
%classification_model_1 = fitcsvm(data, 'Purchased~Age+Es
timatedSalary', 'KernelFunction', 'linear');
```

```
%classification_model_1 = fitcsvm(data, 'Purchased~Age+Es
timatedSalary', 'KernelFunction', 'gaussian');
%classification_model_1 = fitcsvm(data, 'Purchased~Age+Es
timatedSalary', 'OutlierFraction', 0.1);
%%--------------Test and Trainsets----------------------------
%--------------------------Code--------------------------
cv = cvpartition(classification_model.NumObservations,
'HoldOut', 0.2);
cross_validated_model = crossval(classification_model,
'cvpartition', cv);
cross_validated_model_1 = crossval(classification_
model_1,'cvpartition', cv);
%%--------------Making Predictions for Test sets---------------
%--------------------------Code--------------------------
Predictions = predict(cross_validated_model.Trained{1},
data(test(cv), 1:end-1));
Predictions_1 = predict(cross_validated_model_1.
Trained{1}, data(test(cv), 1:end-1));
%%--------------Analyzingthepredictions--------------------
%--------------------------Code--------------------------
Results = confusionmat(cross_validated_model.Y(test(cv)),
Predictions);
Results_1 = confusionmat(cross_validated_model_1.Y
(test(cv)), Predictions_1);
```

6. Ensemble Classifier on Wearable Sensor Gathered Dataset

personid	Age	sex	hpm	wbpm	tbpm	nbpm	status		
1	70	1	109	78	94	117	1		
2	67	0	160	76	109	117	1 status	Normal/Abnormal	
3	57	1	141	76	80	96	1 hpm	heart beat per min	
4	64	1	105	78	120	114	1		
5	74	0	121	71	120	75	1		
6	65	1	140	77	97	103	1 wbpm	Wristbpm	
7	56	1	142	90	98	83	1 tbpm	thumb bpm	
8	59	1	142	140	84	123	0 nbpm	neck bpm	75 - 128
9	60	1	170	78	108	114	1		
10	63	0	154	97	109	93	1 max heart rate=220-Age		

```
clc
clear all
data = readtable('train.csv');
%%--------------Building Classifier--------------------------
```

```
%--------------------------Code--------------------------
classification_model = fitcensemble(data, 'vulnerable');
cv = cvpartition(classification_model.NumObservations,
'HoldOut', 0.4);
cross_validated_model = crossval(classification_model,
'cvpartition', cv);
Predictions = predict(cross_validated_model.Trained{1},
data(test(cv), 1:end-1))
%
%%%testset to print test set data uncomment below line
code
%data(test(cv), 1:end-1)
%%-------------Analyzing the predictions--------------------
%--------------------------Code--------------------------
confusionmatval = confusionmat(cross_validated_model.Y(-
test(cv)), Predictions);
TP=confusionmatval(1,1);
FN=confusionmatval(1,2);
FP=confusionmatval(2,1);
TN=confusionmatval(2,2);
Accuracy=((TP+TN)/(TP+TN+FP+FN))
```

7. Discriminant_Analysis_Classification on Wearable Sensor Gathered Dataset

```
clc
clear all
%%--------------Importing the dataset----------------------
%--------------------------Code--------------------------
data = readtable('train.csv');
classification_model = fitcdiscr(data, 'vulnerable',
'DiscrimType', 'diagquadratic');
%please define your classifier here
%%-------------Test and Trainsets---------------------------
%--------------------------Code--------------------------
cv = cvpartition(classification_model.NumObservations,
'HoldOut', 0.4);
cross_validated_model = crossval(classification_model,
'cvpartition', cv);
%%-------------Making Predictions for Testsets--------------
%--------------------------Code--------------------------
Predictions = predict(cross_validated_model.Trained{1},
data(test(cv), 1:end-1))
%%%testset to print test set data uncomment below line
code
%data(test(cv), 1:end-1)
```

```
%%--------------Analyzing the predictions--------------------
%--------------------------------Code-----------------------
confusionmatval = confusionmat(cross_validated_model.Y(-
test(cv)), Predictions)
TP=confusionmatval(1,1);
FN=confusionmatval(1,2);
FP=confusionmatval(2,1);
TN=confusionmatval(2,2);
Accuracy=((TP+TN)/(TP+TN+FP+FN))
```

8. Kernel Naive Bayes

```
clc
clear all
%%--------------Importing the dataset-----------------------
%--------------------------------Code-----------------------
data = readtable('train.csv');
classification_model_1 = fitcnb(data, 'vulnerable',
'Distribution', 'kernel');
cv = cvpartition(classification_model_1.NumObservations,
'HoldOut', 0.4);
cross_validated_model_1 = crossval(classification_
model_1,'cvpartition', cv);
Predictions = predict(cross_validated_model_1.Trained{1},
data(test(cv), 1:end-1))
%%%testset to print test set data uncomment below line
code
%data(test(cv), 1:end-1)
confusionmatval = confusionmat(cross_validated_
model_1.Y(test(cv)), Predictions_1);
TP=confusionmatval(1,1);
FN=confusionmatval(1,2);
FP=confusionmatval(2,1);
TN=confusionmatval(2,2);
Accuracy=((TP+TN)/(TP+TN+FP+FN))
```

Index

Printed in the United States
by Baker & Taylor Publisher Services